T0189923

Introduction to Industrial Internet of Things and Industry 4.0

Introduction to Industrial Internet of Things and Industry 4.0

Sudip Misra
Chandana Roy
Anandarup Mukherjee

CRC Press
Taylor & Francis Group
Boca Raton London New York

CRC Press is an imprint of the
Taylor & Francis Group, an **informa** business

First edition published 2021
by CRC Press

6000 Broken Sound Parkway NW, Suite 300, Boca Raton, FL 33487-2742
and by CRC Press

2 Park Square, Milton Park, Abingdon, Oxon, OX14 4RN

© 2021 Taylor & Francis Group, LLC

CRC Press is an imprint of Taylor & Francis Group, LLC

ISBN: 978-0-367-64471-0 (hbk)
ISBN: 978-0-367-89758-1 (pbk)
ISBN: 978-1-003-02090-5 (ebk)

Typeset in Computer Modern font
by KnowledgeWorks Global Ltd.

Dedicated to Our Families

Contents

Foreword

Industry 4.0 is commonly referred to as the fourth industrial revolution. Many modern technological developments such as artificial intelligence, quantum computing, gene sequencing, augmented and virtual reality, 5th generation (5G) wireless communication technologies, and 3D printing acted as the stepping stone toward the evolution of Industry 4.0. Industry 4.0 focuses on the transformation of industrial processes through the integration of modern technologies such as sensors, communication, and processing. Technologies such as Cyber-Physical Systems (CPS), Internet of Things (IoT), Cloud Computing, Machine Learning, 3D printing, and Data Analytics are generally considered as the various arms of Industry 4.0, which are the drivers necessary for the transformation.

Industrial Internet of Things (IIoT) concerns the application of IoT in industries to modify the various existing industrial systems by primarily digitizing them. IIoT links automation systems with enterprise, planning, and product lifecycles. IIoT is a relatively new and emerging paradigm, which interconnects sensors and actuators to enhance industrial working conditions, product quality, machine lifetime, automated fault detection and maintenance, and optimal asset utilization. The involvement of various heterogeneous types and classes of sensors and actuators, and the ensuing communication among them for data and information transfer form the crux of IIoT. In addition to these, the traditional industrial setups can be radically transformed with the integration of advanced technologies such as machine-to-machine (M2M) communication, 3D printing, advanced analytics, nanotechnology, augmented reality (AR), and virtual reality (VR).

Industrial IoT and Industry 4.0 are emerging topics, which have the potential to solve many long-standing, real-life issues associated with industries. Owing to the relative newness of this topic, there are very few books in the market related to Industrial IoT and Industry 4.0, which cover both the technical and business perspectives of these technologies. This book is written in a very lucid and elaborate manner. The use of illustrations, real-life examples, and case studies makes this an ideal resource for both beginners and established readers working on IoT in industries and Industry 4.0. This book covers the overview of IoT, basics of Industry 4.0 and IIoT, and advanced technologies to be integrated with the existing technologies to make the industries Industry 4.0 and IIoT compliant. Some of these are (1) Drivers of Industry 4.0, (2) Smart Business Perspectives, (3) Industrial Internet, (4) Business Models and Reference Architecture of IIoT, (5) Smart Factories, (6)

Lean Manufacturing System, (7) Big data and analytics, (8) AR/VR, and (9) Machine Learning and Data Science.

This book has been planned to cover the important concepts of IIoT in detail, including communication, connectivity, and control, which mostly form the core of IIoT and the knowledge of which is indispensable for architecting an IIoT-based solution. The entire book is presented in a comprehensive manner so that the reader gets a crystal-clear introduction to IIoT and Industry 4.0. Additionally, this book provides the basic background knowledge of the technological and communication aspect of IoT. The book will help the beginners to have a basic idea of Industry 4.0 and IIoT. Personally, I feel that the following features make this book stand out as an excellent learning and teaching resource: (a) Preliminaries and background information on IoT, Industry 4.0 and Industrial IoT, (b) Self-descriptive illustrations help in visualizing complex concepts, (c) Exercises at the end of chapters test the learner's understanding of contents, (d) Conceptual questions at the end of this book test the understanding of learners, (e) Real-life use cases and examples of IoT in real industries motivate new learners to delve deeper into Industry 4.0 and Industrial IoT.

The authors of this book are globally acclaimed researchers, all of whom have published a number of research papers in this domain in highly impactful journals/magazines such as the IEEE Communications Magazine, IEEE Transactions on Parallel and Distributed Systems, IEEE Transactions on Communications, IEEE Transactions on Mobile Computing, IEEE Transactions on Sustainable Computing, IEEE Transactions on Vehicular Technology, IEEE Internet of Things Journal, Elsevier Computer Networks, IEEE Systems Journal, and many more. Sudip is well known in the community for his research achievements in the broad domain of Internet of Things. He has more than a dozen books, which are published by globally renowned publishers such as Cambridge University Press, Wiley, Springer, World Scientific, and CRC Press. Due to his significant research contributions, his works were recognized with different fellowships and awards such as highly prestigious Abdul Kalam Technology Innovation National Fellow (India), the Fellow of the National Academy of Sciences (India), the Fellow of IETE (India), IET (U.K.), RSPH (U.K), IEEE Communications Society Outstanding Young Researcher Award, Humboldt Fellowship (Germany), Faculty Excellence Award (IIT Kharagpur), Canadian Governor General's Academic Medal, NASI Young Scientist Award, IEI Young Engineers Award, SSI Young Systems Scientist Award and so on. He serves as the Editor of IEEE Transactions on Vehicular Technology, and the Associate Editor of IEEE Transactions on Mobile Computing, IEEE Transactions on Sustainable Computing, IEEE Systems Journal, and IEEE Networks. He is also the IEEE Communications Society Distinguished Lecturer for the year 2020 through 2021. Besides Sudip, the other authors, Chandana and Anandarup, are highly awarded and experienced senior researchers in Sudip's Smart Wireless Applications and Networking (SWAN) Lab at IIT Kharagpur. Anandarup, who is presently associated with the University of Cambridge

(U.K) has been awarded the Gandhian Young Innovation Award (GYTI) by the President of India in 2018 for his socially relevant innovation. Along with Sudip, Anandarup also serves as one of the co-founders and directors for their entrepreneurial venture Sensordrops Networks Pvt. Ltd.

Besides being a powerful reference material, the authors have managed to make this book a great teaching aid, which will provide learners with a basic exposure to the transition of traditional industries to Industry 4.0 and Industrial IoT compliant ones. Therefore, this book can be used as a textbook in universities and colleges. As this book is concisely written and accompanied by useful illustrations, it can be easily used as a textbook across multiple disciplines and educational backgrounds. This book is suitable for teaching undergraduate as well as postgraduate courses in Industrial IoT, Industry 4.0, and its applications. Further, this can also act as a good "go-to-book" for research that focuses on communication, connectivity, interoperability in Industry 4.0 and IIoT, and application-specific case studies. Finally, it can also be used as a professional reference with highly probable interest in the key enabling technologies of Industry 4.0 and Industrial IoT in industries, business models and reference architectures necessary to integrate new and legacy technologies in industries.

I wish the book all the success that it deserves!

Dr. Tomohiko Taniguchi, *Fellow of IEEE*,
Fujitsu Laboratories Limited, Japan

Author Biographies

Dr. Sudip Misra is a professor in the Department of Computer Science and Engineering at the Indian Institute of Technology Kharagpur. He is also the Abdul Kalam Technology Innovation National Fellow. He earned a Ph.D. in Computer Science at Carleton University, Ottawa, Canada. His current research interests include algorithm design for emerging communication networks. He is the author of over 300 scholarly research papers, more than 120 of which were published by reputable organizations such as IEEE and ACM. He is elected as *Fellow* of the National Academy of Science India (NASI). He is also the IEEE Communications Society Distinguished Lecturer for 2020–2021.

Dr. Misra's work was recognized at various conferences. He received the IEEE ComSoc Asia Pacific Outstanding Young Researcher Award in 2012. He was also the recipient of several academic awards and fellowships including the Young Scientist Award of the National Academy of Sciences of India, the Young Systems Scientist Award of the Systems Society of India, the Young Engineers Award of the Institution of Engineers, Carleton University's Governor General's Academic Gold Medal awarded to an outstanding graduate student doctoral level, and the Swarna Jayanti Puraskar Golden Jubilee Award of the National Academy of Sciences of India. Dr Misra also received a prestigious NSERC post-doctoral fellowship from the Canadian government and a Humboldt research fellowship in Germany.

Dr. Misra is the editor/associate editor of *IEEE Transactions on Vehicular Technology*, *IEEE Transactions on Mobile Computing*, *IEEE Transactions on Sustainable Computing*, *IEEE Systems Journal*, and *IEEE Network*. Dr. Misra has published 11 books in the areas of wireless ad hoc networks, wireless sensor networks, wireless mesh networks, communication networks and distributed systems, network reliability and fault tolerance, and information and coding theory, published by reputed publishers such as Cambridge University Press, Springer, Wiley, and World Scientific. He has chaired several international conferences and workshops and was invited to deliver keynote lectures at more than 20 conferences in USA, Canada, Europe, Asia, and Africa.

Chandana Roy is a senior researcher in the Department of Industrial and Systems Engineering at the Indian Institute of Technology, Kharagpur, India (IIT Kharagpur). Her research interests include Internet of Things (IoT), Industrial Internet of Things (IIoT), Sensor Cloud, and Wireless Body Area Network (WBAN). She coled the planning, organization, and delivery of the massively popular NPTEL MOOC course on Industrial Internet of Things and Industry 4.0. She is an active member of the Smart Wireless Applications and Networking (SWAN) Lab at IIT Kharagpur. Additionally, she worked as an organizing committee member in various short-term courses and workshops organized at IIT Kharagpur. Chandana received fellowships from the Ministry of Human Resources and Development (MHRD), India, and the Defence Research and Development Organization (DRDO), India. She was earlier an assistant professor in the Department of Electrical Engineering at Aryabhatta Institute of Engineering and Management (AIEMD), Durgapur. Chandana's works are published in reputed conferences such as IEEE ICC, WCNC, and GLOBECOM, and journals. She serves as a peer reviewer in reputed IEEE, Wiley, and Elsevier journals/transactions.

Anandarup Mukherjee is a senior researcher in the Department of Computer Science and Engineering at Indian Institute of Technology, Kharagpur (IIT Kharagpur). He concurrently serves as the Founder and Director of Sensordrops Networks Pvt. Ltd., a govt. recognized startup incubated at the Science and Technology Entrepreneur's Park at IIT Kharagpur. His research interests include, but are not limited to, networked robots, unmanned aerial vehicle swarms and enabling deep learning for these platforms for controls and communications. These solutions are mainly aimed at intelligent and autonomous control

and coordination between ground stations and unmanned aerial vehicles (UAVs) as well as between multiple UAVs in flight.

Anandarup is also associated with the hugely popular NPTEL MOOC course on "Introduction to Internet of Things". He has given talks on many topics at both national and international venues on topics ranging from IoT,

UAVs, and Sensor Networks. He has also been associated with various national and international funded projects of organizations such as ITRA- Media Lab Asia, ICAR, and British Council. Earlier, Anandarup served as an Assistant professor in the Department of Electronics & Communication Engineering at University of Engineering & Management, Jaipur (2012–2014) and as a Lecturer in the Department of Information Technology at the Institute of Engineering & Management, Kolkata (2011–2012).

Anandarup has been awarded various prestigious awards such as the 2018 Gandhian Young Technological Innovation award by the Hon'ble President of India for "Socially Relevant Innovation", the Dr. Amulya K. N. Reddy Award awarded by the Hari-om Ashram Prerit Society for commercialization of prototype, and from organizations such as IBM and Johnson Controls at IIT Kharagpur. He has been popularly covered in national print and televised media for his works on UAVs and IoT based systems in healthcare. Till date, he has a total of more than 30 referred scientific publications in renowned SCI-indexed journals and globally famous conferences. such as IEEE INFOCOM, WCNC, and GLOBECOM. He has more than 120 citations with an h-index of 8 (Google Scholar) till date. He regularly serves as a peer reviewer in various highly impactful IEEE Transactions, and other IEEE, Elsevier, Springer, and Wiley journals.

Preface

Overview and Goals

Industrial IoT (IIoT) and Industry 4.0 are newly developed and fast emerging domains of interest among students, researchers, personnel from academia, and industries. Due to the popular demand of this topic in both the industries and academia, this book is written to serve the diverse readership it aims to attract. Readers from the domains of computer science and engineering, mechanical engineering, information technology, industrial engineering, electronics engineering, and other related branches of engineering and technology are expected to benefit directly from this book. This book is designed to cover the complex concepts of the advanced technologies of IIoT and Industry 4.0 in a very lucid manner. The content of this book exposes the readers to real-life applications alongside the relevant theory.

This book evolved from one of Prof. Misra's popular Massive Open Online Courses (MOOC) on Introduction to Industry 4.0 and IIoT, which are hosted by NPTEL (Govt. of India). This course typically sees student registrations of over 8000 during every run. Almost all the participants ranging from academicians, students (UG and PG), working professionals, business people, hobbyists, and many others have repeatedly identified the lack of a concisely written book on IIoT with sufficient exposure to underlying technologies. This motivated us to write this book to cater to the knowledge requirements of these vast groups of people. Additionally, this book aims to be the first of its kind, which can act as a textbook on the emerging paradigm of Industry 4.0 as well as Industrial IoT. It is also expected to serve casual readers as a friendly guide for those looking into various aspects of Industrial IoT.

Pedagogical Aids

We have included various pedagogical aids to serve the reader of this book to swiftly grasp the contents and the treatment of the various topics. We have provided a set of conceptual questions at the end of this book. For solving these questions, the reader must truly grasp the concepts covered in the different chapters. Additionally, we provide visual presentations of the chapters covered so that it can be used as a teaching aid in colleges and universities. Each chapter of this book has the following pedagogical components:

- *Learning outcomes*, which gives an initial glimpse into the chapter.

- Easy-to-understand and self-descriptive illustrations.

- *Check yourself* exercises, which encourages readers to explore additional topics on their own. This will gradually enable the readers to easily find terminologies, technologies, and concepts on their own as they start grasping the more advanced chapters of this book.

- *Summary* provides a concise outcome for each chapter.

- *Exercises* allows the readers to self-access their knowledge and also enables them to test their understanding of the concepts/technologies covered.

Subject Area

This book spans the interest domains of Internet of Things (IoT), IIoT, Industry 4.0, Fog Computing, Cloud Computing, Industrial Revolution, Smart Factory, Industrial Internet, Business models, IIoT Reference Architecture, Analytics, Communication technologies, Interoperability, Use cases.

Target Audience

This book primarily targets the undergraduate and postgraduate technical readers who are either looking to delve into the emerging domain of Industrial IoT or are taking Industry 4.0 targeted introductory as well as advanced courses. The book is additionally designed to address the curious, non-technical reader by enabling them to understand the necessary concepts and terminologies associated with IIoT and Industry 4.0. The multifaceted coverage of this book can be used as a quick reference for covering the introductory concepts and challenges in IIoT and Industry 4.0, which may be of significant use to working professionals, academicians as well as researchers across various industries.

Organization and Features

Industry 4.0 concerns the transformation of industrial processes through the integration of modern technologies such as sensors, communication, and computational processing. Technologies such as Cyber Physical Systems (CPS), Internet of Things (IoT), Cloud Computing, Machine Learning, 3D printing, and Data Analytics are considered to be some of the drivers necessary for the transformation of traditional industries to the modern ones. Industrial IoT focuses on industry-specific applications of IoT to improve existing processes and systems. IIoT links the automation system with enterprise, planning, and product lifecycle.

Industry 4.0 is also known as the fourth industrial revolution. The integration of various technologies such as artificial intelligence, quantum computing, gene sequencing, augmented and virtual reality, 5th generation (5G) and beyond wireless communication technologies, and 3D printing formed the stepping-stone toward the evolution of Industry 4.0. General Electric (GE) coined the term Industrial Internet, which described industrial transformation of machines through cyber-physical systems, advanced analytics, and cloud-based platforms. Later, the Industrial Internet Consortium (IIC) and GE concluded that IIoT was a synonym for the Industrial Internet.

IIoT applies IoT-based solutions to improve processes and systems in manufacturing, aviation, transportation, supply chain, mining, and healthcare. Typically, IIoT is the integration of Information Technology (IT) and Operational Technology (OT). It is an emerging technology, which interconnects sensors and actuators to enhance the working conditions in industries, improve product qualities, enhance machine lifetime, promote automated fault detection and maintenance, and optimally utilize various assets. Heterogeneous sensors and actuators communicate with one another to transfer data among them and eventually to a central server, in most implementation platforms.

The book covers the significant aspects of IIoT in detail, including sensors, actuators, data transmission, and data acquisition, which mostly form the core of IIoT, and the knowledge of which is indispensable for architecting an IIoT-based solution. The entire book is presented in a comprehensive manner so that the reader gets sufficient clarity of knowledge. The book is expected to help the beginners to have a basic idea of Industry 4.0 and IIoT. Therefore, the first module comprises the overview of IoT – various application-based and infrastructure-based protocols, cloud computing, fog computing, and applications of IoT.

The second module provides the basic knowledge of Industry 4.0 and IIoT. It provides a knowledge of the different phases of development of the industries over the ages and the various impacts of industrial revolution on the environment. CPS integrates physical objects and computer-based algorithms, which communicate among themselves and learn from their locality.

This module also covers the basics of CPS and interrelation between CPS and IIoT. Additionally, the readers are introduced to the concept of industrial Internet, which was introduced by GE. Industrial Internet arose from the combination of physical and digital worlds. Few sections of this module also discuss the applications of Industry 4.0 and IIoT. The various advanced and complex technologies in the industries result in the conversion of the traditional things/devices into intelligent and smart devices. Megatrends and tipping points are the major drivers of Industry 4.0, which are also discussed in brief. As discussed earlier, development in industries initiated with the first industrial revolution. However, to maintain sustainability in the industries, globalization and socio-economic factors play an important role. The various facets related with these factors are briefly discussed in the third chapter. On the other hand, huge volume and variety of data are generated at each time instant from these devices. These data are required to be secured from malicious or unauthorized access. The concept of cybersecurity and necessary constituents are briefly discussed in this module. Therefore, to safeguard the data from attacks, knowledge of various cybersecurity threats is necessary. Part II of the book includes the business models and reference architecture of IIoT. This module also sheds light on the basics of a business model, primary units of the business model, requirement of a business model and types, and reference architecture in IoT. The major components of a business model in IIoT are also discussed in this module. IIoT paves the way toward the creation of new business opportunities. There are two different aspects of business opportunities in IIoT, which are entrepreneurship and transaction. Further, the Industrial Internet Consortium (IIC), an open membership organization, has published the Industrial Internet Connectivity Framework (IICF) to unlock the data in isolated systems. This module also covers the Industrial Internet Reference Architecture (IIRA), a standard open infrastructure, which helps in interoperability, map the applicable technologies, and help in standard development. To measure the performance of the overall organization, Key Performance Indicators (KPIs) are designed for each organization. Based on the outcome, the KPIs are classified as leading or lagging.

Part III presents the technological aspects of Industry 4.0 and IIoT. The development of large number of tools and advanced technologies during the Industry 4.0 is resulting in the widespread usage of IoT in the industries. The primary factor responsible for the evolution of these new technologies are access to the real-time data and feedback provided, based on the analysis of the data. This module comprises the necessary transformations for industrial development. The digital transformation in Industry 4.0 introduces digitization to connect products and control systems, analytics, efficient resource utilization and services. Bulk amount of data is generated from each of the devices in real-time. Therefore, the features of storage and analysis of data, estimation and provision of necessary processing capacity, efficient management of energy consumption, and maintenance of back-ups of the data generated, are important, which is why this separate module has been dedicated to introduce these. To provide computational speed, storage, security, and effectively possess the data, cloud is an obligatory technology in modern IIoT implementation. This module also sheds light on the various industrial cloud and fog providers. A few sections of this module discuss the key features and applications of augmented and virtual reality, which provides an enhanced view of the real-world environment. With the help of head-mounted displays, users get the illusion of immersive experience in the environment. The various components of AR and VR are input devices such as mouse, joystick, tracking balls, head-mounted display, and console or personal computer. Smart factory and its characteristics are also discussed in this module. Smart devices, smart machines, advanced technologies, and data analysis form the components of smart factory. The transformation of industries results in self-optimization of the performance, automatic detection of the abnormality and automated maintenance, and self-adaptation with the environment. Therefore, real-time monitoring and tracking of messages is necessary. Lean manufacturing system is another important aspect of Industry 4.0, which helps to reduce wastes from processes. In respect of Industry 4.0, the requirement, implementation, and impact of lean manufacturing systems are briefly covered in this module.

Part IV discusses the key enabling technologies of IIoT sensors, actuators, data communications, data acquisition, and their various aspects. Based on these aspects, the intelligent devices' data are transmitted to the server. This book illustrates these protocols in a lucid and simple manner, which, we hope, will help both new and seasoned learners. Different types of electromechanical instruments and the associated systems are used in industries to control various industrial units or processes. Further, the various Ethernet- and Fieldbus-based communication protocols such as ModBus, Fieldbus, and HART are discussed in brief. The basics of the industrial control systems and control loops are described in this part of the book. Further, the basics of different types of Industrial control systems (ICS) – Programmable Logic Controllers, Distributed Control Systems, and Supervisory Control and Data Acquisition systems are also described.

Part V discusses the basics of predictive and prescriptive analytics applied in IIoT-based implementations. For new learners, this book describes the basics of machine learning and its various concepts. This book describes how machine learning can help in building IIoT-based systems. The impact of deep learning, and various applications of analytics are also covered in this part. In addition to this, Part VI discusses some of the applications and case studies of IIoT. This book covers various possible applications of IIoT – healthcare, agriculture, chemical industries, plant safety and security, and inventory management. Additionally, this module also sheds some light on some real-life industry case-studies. Each chapter of Parts I–VI enables all types of readers to understand the requirements, communication protocols, processing, control and feedback, advanced technologies, and analytics.

Suggested Use

The book has been designed to be used both as a textbook and a reference book. The exercises at the end of each chapter and the conceptual questions at the end of this book would serve as a good indicators of the undergraduate readers' progress in grasping the concepts covered in the book. Working professionals in other domains and new learners (not familiar with networking and the nuances of computing sciences) can follow the same approach.

Graduate and research students in electronics, computer sciences, electrical engineering, industrial engineering, and other similar domains can use this book directly from the second part, as they would already be aware of the introductory concepts. Working professionals in the allied domains of electronics and computer sciences can also follow the same usage for this book as the graduate students. This book serves as a ready reference for the numerous topics, which are commonly encountered in designing IoT-based solutions.

Finally, the curious reader who aims to work on Industry 4.0 and Industrial IoT, without having any prior knowledge of this domain and computer sciences, should first explore the sixth part of this book, which covers various real-life case studies of Industrial IoT in various domains. This will help the reader understand the usability of Industrial IoT in the context of the unique challenges faced for each domain.

Endorsements

"This book is an excellent resource for professionals working on IoT in the industry domain and/or are planning to work with various aspects of Industrial IoT. It seamlessly balances the technical and business aspects of industrial IoT and Industry 4.0."

Dr. Veena Mendiratta, *Senior Member, IEEE*
Network Reliability and Analytics Research Lead,
Bell Labs., Nokia, USA

"Introduction to Industry 4.0 and the Industrial Internet of Things' is a highly-readable and accessible introduction to this vast and complex area. With this text students or industrial practitioners can obtain a deeper understanding of the foundational elements, technical challenges, and practical solutions in the space of IIoT."

Dr. Giridhar D. Mandyam,
Chief IoT Security Architect,
Qualcomm Inc., USA

"The Industrial Internet of Things and Industry 4.0 are strongly interdependent and driven by new and game changing digital paradigms. The book covers Industrial IoT and Industry 4.0 in an in-depth and lucid manner. Both the topics are of paramount interest and importance to industry and academia today.

The authors need to be congratulated for coming out with a unique and comprehensive book that covers a wide range of topics that include business models, applications, key enabling technologies and case studies across diverse industries. The book should serve extremely useful for both practitioners and researchers working on various dimensions of the digital ecosystem of which Industry 4.0 is a key element. "

Dr. Rajeev Shorey, *Fellow of National Academy of Engineering*
Chief Executive Officer,
The University of Queensland - IIT Delhi Academy of Research
(UQIDAR), Indian Institute of Technology Delhi, India

"This is an outstanding book for students, academicians and industrialists who need to understand the complexities of both IIOT and Industry 4.0 and it potential benefits. The authors have compiled a comprehensive, clear and cogent treatise of these subjects that will help any reader gain deeper understanding in these critical/fast changing areas. A much needed contribution to the literature that fills a critical gap in the shared knowledge on the subject. I"must" read."

Dr. Jeffrey D. Tew
Chief Scientist,
Manufacturing Research Lab., Tata Consultancy Services, USA

Pre-requisites

1

Overview of Internet of Things

1.1 Learning Outcomes

■ This chapter covers the basics of the Internet of Things (IoT), architecture, and its protocols.

■ New readers will be able to have an understanding of the basic protocols of IoT, which will help them to grasp the concepts of the Industrial Internet of Things (IIoT).

■ Additionally, the readers will get a flavor of cloud computing, fog computing, and big data technology.

■ For seasoned learners, this chapter will ease the revision of their exixting knowledge.

■ The reasons for the shift of technology from the traditional methods to cloud, mobile cloud, and fog computing are outlined.

■ The concept of providing physical sensor nodes as service on-demand is also outlined in this chapter.

1.2 Introduction

The rapid advancement in technologies has resulted in automation, and real-time analytics being applied across various interesting fields such as wildlife monitoring, agriculture, military, healthcare, manufacturing, transportation, supply chain, and inventory management. Internet connectivity has become a basic necessity around the globe where IoT connects billions of people by mobile devices, giving rise to immense processing power, storage capabilities, and knowledge access throughout the world. The interconnection of the physical devices such as appliances, vehicles, and human beings with the help of sensors, actuators, and software leads to the formation of a *network*. This technology enables the physical devices to communicate, interact, process, and exchange data among themselves. As shown in Fig. 1.1, physical objects such as mobile, laptop, bicycle, car, desktops, and even human beings can coordinate and communicate among themselves to form a network. The network of these physical devices is termed as '*Internet of Things*' (IoT). *Kevin Ashton*, a British Technology pioneer, at the Massachusetts Institute of Technology (MIT), coined the term '*Internet of Things*.' He used the term IoT to explain a system, which connects the physical devices over the Internet through sensors. However, CISCO estimates that the concept of IoT evolved between 2008 and 2009. With the help of IoT, physical things are converted into smart objects, in turn enabling them to be remotely monitored and controlled over the network. Therefore, this leads to amelioration in efficiency, automation, accuracy, and advanced applications.

There has been a significant impact of IoT on the growth of connected devices in the past few years. Although IoT has intricate functionalities, its internal working is quite complicated because of the multiple sensing and communication protocols involved with it. IoT has a wide range of applications in the field of agriculture, healthcare, manufacturing, transportation, environmental monitoring, and others. There are few issues in IoT, which require addressing, such as interoperability among the devices, scalability, and processing of the colossal amount of data generated.

This chapter mainly addresses the basics of IoT, the architecture of IoT, communication protocols of IoT, basics of networking, cloud computing, fog computing, and big data. The readers are suggested to go through the suggestions given in the *CHECK YOURSELF* section.

> Check yourself
>
> Networking: Basic concepts, OSI model, Layers in the OSI model, TCP/IP Protocol Suite, Physical and Logical Addressing.

1.3 IoT Architecture

The IoT platform connects heterogeneous devices or systems over the network. A Service-Oriented Architecture (SOA) can be designed to validate the basic IoT architecture. There are four different layers of IoT — (1) sensing layer, (2) networking layer, (3) service layer, and (4) interface layer, as shown in Fig. 1.2 [6].

(a) *Sensing layer*: The sensing layer consists of IoT devices that are equipped with *sensor nodes*, Bluetooth devices, scalar sensors, analog sensors,

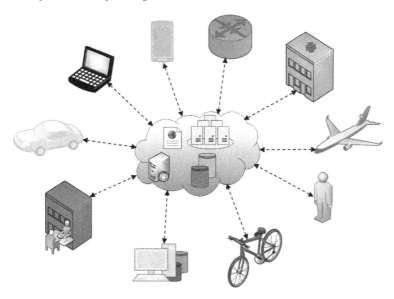

FIGURE 1.1: Internet of Things.

digital sensors, and *RFID tags* as outlined in Fig. 1.2. The sensor nodes sense, process the sensed information, transmit the real-time information, and communicate among themselves. Moreover, these sensor nodes consume low power and require low data-rate connectivity. They form a Wireless Sensor Network (WSN) among themselves. Based on the application type, these sensor nodes are grouped. Each of the IoT devices has a *Universal Unique Identifier* (UUID).

(b) ***Networking layer***: The networking layer helps the IoT devices to share information with other devices. Additionally, this layer handles the colossal amount of data generated by these IoT devices. Certain QoS requirements should be fulfilled to retain communication among these heterogeneous devices. Concerning Fig. 1.2, the networking layer consists of Social Networks, Mobile Networks, WLAN, Internet, Databases, and WSNs. Therefore, the design issues to be kept under primary considerations are latency, scalability, bandwidth requirements, energy efficiency, security, and privacy.

(c) ***Service layer***: The service layer consistently integrates the services and applications in IoT. This layer executes the overall workflow process, which includes information exchanges, communication, storage, and data management. Additionally, various forms of predictive analytics are performed in this service layer. Further, this layer also maintains trust and utilizes the information extended by the other services.

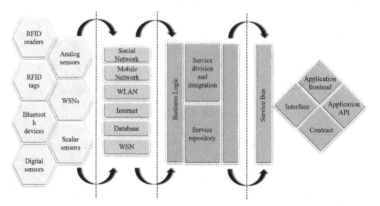

FIGURE 1.2: IoT architecture.

(d) **Interface layer:** The heterogeneous IoT devices do not always follow the same IoT protocol. Due to the presence of various types of IoT devices, problems in interaction exist among these devices. In addition to this, with the rapid increase in IoT devices, it becomes quite difficult to connect, communicate, and operate these devices dynamically. Therefore, an interface layer is essential, which can systematically streamline the management and interconnection among things. For example, a call center operator who can communicate only in English and Chinese receives a call in Spanish. If the received call is not interpreted dynamically, the operator will fail to make sense of the ensuing communication. Therefore, there must be a common language for communication between them.

1.4   Application-based IoT Protocols

There are few standard protocols involved for various applications in IoT, as illustrated in Fig. 1.3. As the network user can be a human being, a machine, or an object, there are few complications involved in the IoT networks [7], which are as follows:

(a) Rapid growth in the number and heterogeneous IoT devices.

(b) Management of the IoT devices.

(c) Standardization of protocols within the network.

Application-based IoT Protocols		
	Infrastructure	6LowPAN, IPv4/IPv6, RPL
	Identification	EPC, IPv6, uCode
	Communication	Wi-Fi, Bluetooth, LPWAN
	Discovery	Physical Web, mDNS, DNS-SD
	Data Protocols	MQTT, CoAP, AMQP, Websocket
	Device Management	TR-069, OMA-DM
	Semantic	JSON-LD, Web Thing Model
	Multi-layer Frameworks	Alljoyn, Weave, Homekit

(a) Various application-based IoT protocols.

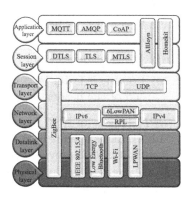

(b) OSI model-based IoT protocols.

FIGURE 1.3: Different IoT protocols.

Check yourself

IEEE Standards, Ethernet, IPv4 addressing, IPv6 addressing, Router, Switch, Gateway, Unicast routing, and Multicast routing.

1.4.1 Infrastructure-based protocols

There are certain infrastructure-based IoT protocols such as 6LowPAN, IPv4/IPv6, and RPL, as depicted in Fig. 1.3(a).

(a) *6LowPAN*: This is a low-power Internet Protocol version 6 (IPv6)-based technology for packet delivery in Wireless Personal Area Networks. 6Low-PAN allows low-power devices with limited processing power to be connected over the Internet. It supports both 16-bit short and IEEE 64-bit addresses. Further, 6LowPAN is capable of handling a large number of devices deployed in the ad-hoc form. The main application areas of 6Low-PAN are low-power radio communication technologies, smart grid, and M2M applications [13].

(b) *IPv6*: IPv6 is the successive version of Internet Protocol version 4 (IPv4), which deals with addressing over IP-based networks. The exponential growth of the Internet due to technological advances has saturated the existing address space [14]. IPv6 fulfills these shortcomings of IPv4. The features of IPv6 are as follows:

(a) It provides a larger address space.

(b) The header is represented in a simplified format.

    (c) Every system has a unique identification code, enabling the implementation of end-to-end connectivity.

    (d) The host devices can be auto-configured.

    (e) Routing is faster because the unnecessary information is placed at the end of the header.

(c) *RPL*: This is an IPv6 routing protocol for low-power and lossy networks (LLN). LLNs operate on networks with a high loss rate, low data-rate, and instability. RPL provides the routing procedure from multipoint-to-point traffic,[*] as well as point-to-multipoint traffic,[†] are supported [15].

1.4.2 Data protocols

Like the traditional *Open Systems Interconnection* (OSI) model, IoT also has some standard protocols such as MQTT, CoAP, AMQP, WebSocket, and Node. The comparison among the various data protocols are discussed in Table 1.1.

(a) *Message Queuing Telemetry Transport* (MQTT): MQTT is an ISO standard (ISO/IEC PRF 20922) *publish–subscribe* based application layer protocol, which is used in concurrence with the TCP/IP protocol. It is primarily used for making remote connections that require limited bandwidth [9]. Additionally, MQTT is preferred for mobile applications because of its small size, minimal usage of power, and lower requirement of data packets. Secure MQTT – an augmented version of MQTT and MQTT-SN (MQTT for Sensor Networks) – uses lightweight *Key/Ciphertext Policy-Attribute based Encryption* (KP/CP-ABE) to secure the IoT devices [8].

(b) *Constrained Access Protocol* (CoAP): CoAP is a special application-layer protocol used for resource–constrained devices and constrained networks. CoAP is a generic web protocol, which is suitable for the Machine-to-Machine (M2M) communication paradigm. This application layer protocol follows the request–response interaction model between the application points. Additionally, CoAP fulfills the various network requirements such as multicast support, low overhead, simple proxy, and caching abilities. The *Web-based Application Programming Interface* (API) depends on the fundamental Representational State Transfer (REST) architecture. The CoAP protocol recognizes the RESTful environment, for simple HTTP interfaces [10].

---

[*]Multipoint-to-Point Traffic: Traffic flows from devices inside the LLN to a central point.
[†]Point-to-Multipoint Traffic: Traffic flows from a central point to the devices inside the LLN.

(c) *Advanced Message Queuing Protocol* (AMQP): AMQP is an open standard protocol (ISO/IEC 19464) utilized for the transfer of business messages between the organizations, technologies, or business process. The primary characteristics of AMQP are security, interoperability, reliability, queuing, and routing. AMQP processes and delivers messages to multiple customers at the same time instant. In the case of lightweight applications, AMQP permits the complete message transfer from login to logout [11].

(d) *WebSocket*: WebSocket is an advanced technology which helps in an open, interactive communication session between the client and the server. It establishes a full-duplex single socket connection over which the messages are transmitted between the client and the server. WebSocket protocol has two versions—WebSocket and WebSocketSecure. Firefox, Google Chrome 14, and Internet Explorer possess a secure version of the WebSocket protocol [12].

1.4.3 Transport protocols

(a) *Wi-Fi*: The Wi-Fi protocol is based on IEEE 802.11 standards, which uses radio waves for communication. The term Wi-Fi means *Wireless Fidelity*. It works on half-duplex mode, where the Wi-Fi stations transmit and receive over the same channel. The communication range of Wi-Fi is restricted to a small region, mostly limited to Local Area Networks (LANs). It works in the frequency range of $2.4 - 5$ GHz. No wired connection exists between the sender and the receiver in Wi-Fi technology [17].

(b) *Bluetooth*: Bluetooth is another wireless technology, which is used to connect users in a wireless personal area network (WPAN). It is primarily designed for secure, low-cost, and low-power applications. Bluetooth is standardized as IEEE 802.15.1. It works in the frequency range of 2.4 GHz. The Bluetooth devices are paired with one another through a one-time process called *pairing* [16]. Multiple steps, such as inquiry, paging, and connection, are involved in establishing a connection between two devices.

1.5 Cloud Computing

The term *cloud* refers to the complete infrastructure, services, and software applications, which are dynamically used to perform end-users' tasks. As per Vaquero *et al.* [25], the cloud is defined as, *Clouds are a large pool of easily usable and accessible virtualized resources (such as hardware, development platforms and/or services). These resources can be dynamically reconfigured to adjust to a variable load (scale), also allowing for optimum resource utilization.*

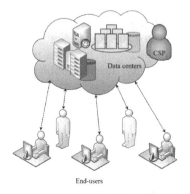

(a) Architecture of cloud.

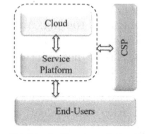

(b) Block diagram of cloud.

FIGURE 1.4: Network architecture: cloud computing.

Features	MQTT	AMQP	Websocket	CoAP
Sponsor	MQTT.org	OASIS	IETF	IETF
Comm- unica-tion Pattern	Publish/ Subscribe	Publish/ Subscribe	Publish/ Subscribe	Request/ Response
QoS level	High	High	Low	Moderate
Power Con-sumption	Low	Medium	Low	Medium
Application	To acquire data from multiple de-vices and transfer it to the infras-tructure	Supports reliable com-munication via message delivery	Helps in interactive communica-tion between the client and the server	Used between devices on the same constrained network
Network Layer	Application layer	Application layer	Application layer	Application layer
Security	SSL	SSL	SSL/TLS	DTLS
UDP/TCP	TCP	TCP	TCP	UDP

Table 1.1: Comparison among the data protocols [20, 18].

This pool of resources is typically exploited by a pay-per-use model in which guarantees are offered by the Infrastructure Provider using customized Service Level Agreements (SLAs) [29].

The various actors of the cloud computing infrastructure are as follows:

(a) End-users: The end-users are the consumers who purchase the cloud ser-vices from the Cloud Service Provider (CSP), as per their requirement.

(b) Cloud Service Provider (CSP): A CSP maintains, upgrades, and manages the pricing of the whole infrastructure of the cloud. For example, Amazon and Google.

(c) Enablers: Enablers are the organizations that act as the seller and deliver the cloud computing products and services. For example, Capgemini, RightScale, and Vordel.

(d) Regulators: Regulators are the government body or organization, which penetrates across all the actors of the cloud infrastructure.

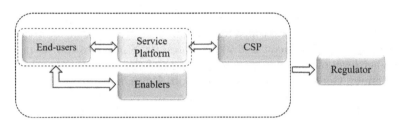

FIGURE 1.5: Work flow among the actors in cloud [26].

The cloud architecture examines the data centers regularly to fulfill the end-users' requirements. Thus, the cloud can be represented as the set of interconnected and virtualized devices, which dynamically provides on-demand services, as shown in Fig. 1.4(a). Additionally, the cloud also reduces IT management complexity, depending on the SLA among the CSP and end-users (Table 1.1). Cloud Computing is the technology that enables the remote provision of on-demand services from a pool of resources on *pay-per-use* basis. There are various characteristics of cloud, which lead to uniqueness and advanced computing technologies in the cloud, as depicted in Fig. 1.5.

(a) The end-users can use the cloud at any time instant, independent of their geographical location.

(b) The end-users have to pay CSP only for the resources used by them (*pay-per-use*).

(c) The resources in the cloud are available to the user dynamically on an on-demand basis.

(d) Infinite resources are available in the cloud, which are dynamically consumed and released by the end-users, as per their requirement.

(e) Cloud allows a large number of end-users to connect to a CSP at the same time.

1.5.1   Types of cloud

The types of cloud are categorized depending upon the services provided by the CSP to the end-users and the infrastructure of the cloud [19]. Fig. 1.6 illustrates the various types of clouds. Based on the services provided by the cloud to the end-users, it is classified into three types, which are as follows:

(a) Infrastructure-as-a-service (IaaS): The hardware or infrastructure such as storage disks, networks, operating systems, and virtual servers are provided access by the CSP to the end-users. IaaS provides end-users with the highest level of flexibility and control over the resources requested by them. For example, Amazon cloud services, Rackspace, and Flexiscale.

(b) Platform-as-a-service (PaaS): A customized development platform is provided to the end-users, which includes an operating system, database, programming environment, and web servers. PaaS provides end-users with a higher level of clarity in the management of the resources than IaaS. For example, Google App Engine, Microsoft Azure, and Engine Yard.

(c) Software-as-a-service (SaaS): The end-users obtain access to the software applications on a *pay-per-use* basis. SaaS is most convenient for the end-users for managing their resources compared to IaaS and PaaS. For example, Google Docs, Gmail, and Office 365.

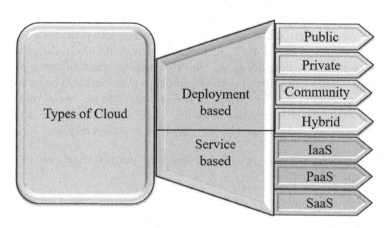

FIGURE 1.6: Types of cloud.

Depending on the deployment of the cloud, it is classified into four types, which are as follows:

(a) Public Cloud: The complete infrastructure is situated at the venue of the concerned organization. The resources of the cloud are publicly shared among the end-users. However, security is an issue in the public cloud because of the shared resources. The end-users do not have any control over the infrastructure of the cloud. Example: Google Apps.

(b) Private Cloud: The cloud infrastructure is completely privately owned by an end-user or organization. The security levels are quite high in a private cloud. However, the services provided to the end-users are expensive, as the end-users have physical control over the infrastructure. Example: Google cloud product certified by the Federal Information Security Management Act (FISMA).

(c) Community Cloud: The possession of the cloud infrastructure is distributed among multiple organizations, which form a community. Each of the members belonging to the community shares similar security and privacy concerns. Example: Terremark's Enterprise Cloud platform in the US.

(d) Hybrid Cloud: The hybrid cloud is a combination of the features of public and private clouds. The cloud is deployed between the concerned organization and the existing infrastructure. Therefore, the resources are managed and given to the end-users, both internally and externally. Example: Hybrid cloud provided by Azure and IBM.

1.5.2 Business aspects of cloud

Cloud computing is a paradigm shift from the traditional way of utilizing IT resources. The cloud computing model has simplified the work of CSP in terms of upgradation, testing of the software, deployment services, and scale applications. Since the infrastructure is deployed at the premises of the service provider, the CSP can remotely monitor and access these resources. Further, cloud computing is a *pay-per-use* business model. The end-users are charged only for the computational resources, storage, and duration of resources used by them. Therefore, the end-users request for the services, as per their requirement. Multiple actors are involved in the cloud infrastructure. Thus, there exist continuous cash inflow and outflow among the different actors in the infrastructure. This abstracts the technical environment of the cloud to be shifted to a broader business perspective of the cloud infrastructure. The key advantages of the business perspective of cloud [24, 26] are as follows:

(a) The charges incurred in buying and maintenance of hardware and software, data centers, power consumption, and cooling are reduced to a great extent using cloud computing.

(b) The cloud computing services are provided to the customers on demand. Therefore, configuration, installation, and software improvements are done by the CSP remotely.

(c) The resources are allocated to the end-users globally. As per their requirement, the economies scale up or down. Therefore, it provides more elasticity to the CSPs in handling the resources.

(d) As the cloud computing services manage the resources securely and remotely with the latest upgraded software, it reduces the latency of the network. Thus, the overall economy of the infrastructure improves.

(e) Cloud computing provides end-users instant access to the infrastructure in various ways. The end-users are completely disconnected from one another, which further increases the flexibility of the infrastructure. Moreover, with the increase in demand in the system, the system gets more balanced, leading to better scaling of the economy.

(f) Cloud computing also provides wide range of applications and services to the end-users, which are: (a) mobile interactive applications that respond in real-time to the information provided by end-users and sensor nodes, (b) parallel batch processing permits huge amounts of processing power to the end-users, (c) business analytics utilizes vast amounts of data to recognize the characteristics of human users, and (d) adds computer-intensive desktop applications which are offloaded to the cloud.

1.5.3   Virtualization: Key aspect of cloud computing

The term *virtualization* indicates the procedure of creating a logical abstraction of the existing physical resources, such as computer network, storage devices, and hardware platforms [49, 28]. The virtual environment created can be a single instance or a coalition of instances of the operating systems, which runs independently on the host machine without intervening in the other programs of the user. Therefore, virtualization helps to run the same application on multiple virtual machines (VM) at the same time. On the other hand, this leads to an increase in the utilization and flexibility of hardware. The benefits of virtualization are as follows:

(a) Requirement of unique hardware features is minimized.

(b) Increase in efficiency, productivity, agility, and responsiveness.

(c) Efficient allocation of the resources.

(d) Reduced chances of data loss due to multiple backups.

(e) Increase in scalability and remote access to the infrastructure.

(f) Pay-per-use model.

The virtually created environment on the host machine remains logically separated from the hardware. The VM (also known as the guest machine) and the host machine are logically linked using a software named as *Hypervisor*. Hypervisor distributes the hardware resources and allows multiple VM instances to run on the same machine. Based on the type of application used and hardware utilization, there are different types of virtualization which are as follows:

(a) Hardware Virtualization: The physical hardware segment is logically divided into multiple logical hardware segments and servers. The resource allocation is done with the help of a hypervisor.

(b) Software Virtualization: Virtual environment is created on the operating system of the host machine.

(c) Storage Virtualization: Different physical storage sources are grouped and utilized as a single repository. For example, performing a hard drive partition.

(d) Application Virtualization: The end-user can remotely access the application running on the host machine through application virtualization.

(e) Desktop Virtualization: Desktop virtualization allows the end-user to access their desktop from any device/location remotely. The end-user usually stores his/her operating system on a server.

(f) Network Virtualization: Network virtualization creates multiple subnetworks on the same physical network, which may/may not communicate among themselves. In case of unavailability of any of the network, the other sub-networks remain unaffected. This increases the reliability of the network.

1.5.4 Mobile cloud computing

Presently, Mobile Cloud Computing (MCC) has emerged as a popular technology, which finds applications in mobile devices. MCC is an integration of both cloud and mobile computing. MCC is described as an infrastructure, where the data processing and storage is done outside the mobile device to the centralized cloud platform [30]. There are extensive applications of MCC in the field of mobile commerce, mobile healthcare, mobile gaming, mobile banking, and mobile government services as illustrated in Fig. 1.7. The advantages of MCC are as follows:

(a) Energy is one of the important constraints of sensor networks. The network lifetime can be extended by applying computation offloading techniques.[‡]

[‡]Computation offloading refers to the transfer of complex computational tasks to the remote server. Thus, offloading tasks lead to a reduction in power consumption and the delay incurred in data dissemination.

(b) The storage capability and processing power of the devices improve. MCC provides end-users the platform to store huge amount of data in the cloud. Therefore, MCC permits access to data from different geographical locations at any time instant.

(c) Due to the enormous increase in the volume of service usage, the service cost has minimized.

(d) Although the end-users demand fluctuates with time, MCC provides the flexibility of rapid deployment of mobile applications to fulfill the demand.

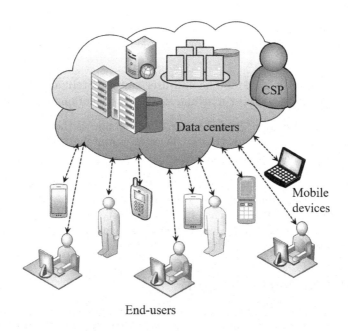

FIGURE 1.7: Mobile cloud computing.

There are certain drawbacks of MCC, which are as follows:

(a) Data security and privacy are the biggest threats of MCC. In case the mobile applications are not properly secured, it may pose threats to the end-users' data.

(b) Quality of Service (QoS) in terms of bandwidth and service availability is another matter of important concern in MCC networks.

(c) On-demand service provisioning to multiple, heterogeneous mobile users is also quite challenging in MCC.

1.6 Fog Computing

The IoT technologies are primarily dependent on the cloud computing services for processing, analysis, and storage of data. On the other hand, billions of Internet-connected IoT devices produce huge amounts of data every second. Thus, the data volume, variety, and velocity cause service latency and poor QoS. *Fog computing* led to the minimization of the latency, improvement in the network bandwidth, and reduces the security issues of the network in real-time [22, 32, 31]. Cloud computing is a centralized framework, where the resources are distributed at the core of the network. On the other hand, fog computing has a decentralized architecture, where the fog nodes are capable of communicating among themselves. Fig. 1.8 describes the decentralized architecture of fog and the block diagrammatic representation of fog. Table 1.2 cites the differences between the cloud and the fog computing. The various features of fog computing are as follows:

(a) Fog nodes are widely distributed over the network at various locations to provide real-time applications.

(b) Fog nodes are heterogeneous and are available in large numbers due to extensive geographic distribution.

(c) The fog nodes have transient storage capability, up to 1–2 hours.

(d) As the fog nodes are placed closer to the source device, it provides a quicker response to the computation, resources, storage, and service request of the end-users.

(e) The fog nodes run the IoT-enabled applications with a response time varying from milliseconds to seconds.

(f) Further, the fog nodes also support the interoperability of services among the various service providers.

As discussed by Bonomi *et al.* [32], fog architecture consists of the *abstraction layer* and *orchestration layer*. The fog abstraction layer supports heterogeneous devices, manages the resources, and controls them. With the assistance of virtualization, this layer provides service to multiple end-users at the same time instant. Additionally, this layer also helps the service providers to keep track of the usage of physical resources by each end-user. On the other hand, the fog orchestration layer provides a dynamic infrastructure, which supports multiple applications. The distributed storage of the resources maintains a high transaction rate update. Further, *Foglet*, a software agent, manages the entire distributed framework of the fog nodes. Additionally, the Foglet agent administers the overall resources and services running on the nodes via the abstraction layer, as depicted in Fig. 1.9.

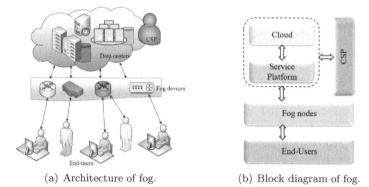

(a) Architecture of fog.   (b) Block diagram of fog.

FIGURE 1.8: Network architecture: fog computing.

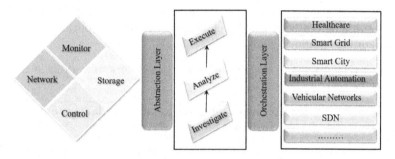

FIGURE 1.9: Architecture of fog layer.

1.6.1   Applications of Fog computing

Fog computing has widespread applications in real-life scenarios such as smart grid, vehicular networks, augmented reality (AR), real-time video streaming, smart city, and software-defined networks (SDN).

Smart Grid: The fog nodes near the grid and sensor nodes process the data generated locally. The primarily processed data is transmitted to the higher tier for storage. In a decentralized micro-grid, fog devices help in reducing the communication overhead. The end-users can optimally utilize the power as per their requirement [33].

Augmented Reality: AR applications require high computation power for processing videos in real-time. Any delay incurred in processing this information is undesirable, as it ruins the user experience [33].

Connected Vehicular Network: The fog devices support real-time communication among the vehicles and between the vehicles and road-side units (RSUs), mobility, and heterogeneity. The safety of pedestrians, drivers, and vehicle owners can be improved through the local processing of the sensed data in almost real-time. An alert is transmitted to the vehicles if the speed

Table 1.2: Comparison between cloud computing and fog computing [102, 22].

Features	Cloud Computing	Fog Computing
Service Location	Within the Internet	Edge of the network
Geographical distribution	Centralized	Distributed
Response time	Minutes	Milliseconds to seconds
Data storage	Long duration	Transient/Short duration
Geographic coverage	Global	Local
Mobility support	Limited	Supported

of the vehicle is detected to be above the safe limit. Similarly, pedestrians will also receive an alert regarding the traffic signals. Accidents, collisions, and on-road congestion are reduced.

1.7 Sensor Cloud

With the emergence of WSNs, the number of physical sensor nodes has globally increased in different applications. However, these applications are single user-centric, where the end-user is the owner of the sensor nodes. To serve and share these applications among multiple end-users or organizations, the concept of the new infrastructure, termed as *sensor cloud*, was proposed [34]. Through the process of *virtualization*, the end-user remains unaware of the actual location of the sensor nodes. Therefore, the physical sensor nodes are provided as service, *Sensors-as-a-Service* (Se-aaS), to the end-users, as per their requirement. As shown in Fig. 1.10, there are three main actors—sensor owner, end-user, and sensor cloud service provider (SCSP) in the sensor cloud infrastructure. The sensor owner deploys the sensor nodes in a particular geographical location and rents their nodes to the SCSP, as shown in Fig. 1.10. Each sensor owner registers their sensor nodes to the sensor cloud infrastructure. An SCSP is a centralized entity, who administers and manages the entire infrastructure.

End-user requests for service/s, by selecting specific templates through a Web portal. Based on their requests, the virtual sensor groups are formed, and the application is served to the end-user. A virtual sensor is formed from multiple physical sensor nodes of a similar type. Therefore, any physical sensor node can be a part of many virtual sensor nodes. The virtual sensor group is formed from multiple virtual sensor nodes. As in cloud computing, the end-user has to pay the SCSP for the services requested by them on pay-as-per their usage basis. Therefore, the end-users are relieved from the costs related to the setup, maintenance, and management of the sensor nodes [35].

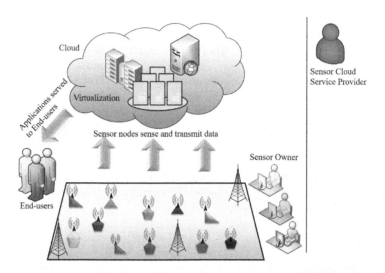

FIGURE 1.10: Sensor cloud infrastructure.

1.7.1 Applications of Sensor Cloud

There is a wide range of physical resource allocation and applications of sensor cloud in the field of target tracking, weather monitoring, hospital services, and military applications.

Target Tracking: In a WSN, the owner deploys the sensor nodes to track the target. The WSN owners of the sensor node refuse to share the information. However, in a sensor cloud-enabled scenario, the organization uses the same physical sensor nodes to provide services on demand. The service provider provides the real-time processing and manages the physical sensor nodes deployed at different geographical locations.

Weather Services: The end-users may request for weather services such as temperature, rainfall, and humidity to the SCSP. Based on the applications requested by the end-user, the virtual sensor groups are created by the service provider. Using virtualization, the same physical sensor node can provide data to multiple end-users at the same time instant. End-users also enjoy the service on a rental basis as per their requirement.

1.8   Big Data

The term Big Data is used to describe the huge volume of data generated from millions of IoT devices in structured,[§] semistructured,[¶] and unstructured[‖] format. Big data provides various opportunities to perform data-enabled decisions. Further, big data is generally characterized by velocity, veracity, volume, variability, value, visualization, and variety.

(a) *Volume*: Volume indicates the huge quantity of the data generated and collected from sensor nodes. Example: Akamai analyses 75 million events a day to target online ads.

(b) *Variety*: Variety indicates the format of the various types of data. Data are mostly present in an unstructured or semistructured format. Example: text, images, audio, and video.

(c) *Velocity*: Velocity indicates the speed of the data generated and processing the data in real-time. Example: 140 million tweets per day on average.

(d) *Veracity*: Veracity denotes the noise and deformation present in the data. The quality of big data also refers to the veracity of the data. The quality of the data depends on the accuracy of the data and the reliability of the data source.

(e) *Value*: Value is the main objective, outcome, prioritization, overall value, and importance given to the data by the stakeholders. Moreover, value incorporates the volume and variety of data. The value of the data is easily accessible, and constant analysis helps to improve the efficiency of the machine, product, or performance of the business.

(f) *Variability*: Variability indicates the unpredictable variation in the data. With reference to the data, meaningful information can be derived from the data. Example: Hashtags, Multimedia, and Geo-spatial data.

(g) *Visualization*: Visualization denotes the representation of the data in pictorial or graphical format. Due to the variety, velocity, and corresponding relationship between the data, it is not very easy to visualize the data.

[§]Structured data are organized in a structured format, stored in databases, and managed using Structured Query Language (SQL). About 20% of the global data available are of structured format.

[¶]Semistructured data is the format of data, which lies between structured and unstructured data.

[‖]Unstructured data are not organized in any pre-defined format. Traditional Relational Data Management system cannot store the data. Around 80% of the global data available are of an unstructured type. Example: satellite images, scientific data, mobile data, and the data obtained from meteorological profiles.

To handle these massive amounts of data, a software framework, *Hadoop*, is used for distributed processing. This software provides the platform for storing the data, management of data in both structured and unstructured formats. Moreover, Hadoop technology allows end-users to collect, process, and analyze data locally.

Summary

This chapter covers the basics of IoT—architecture and application-based IoT protocols.

(a) The interconnection of physical devices, such that they communicate, interact, process, and exchange data among themselves, using a unique identity, is called *Internet of Things*.

(b) The IoT architecture consists four layers—Sensing, Networking, Service, and Interface.

(c) There are different IoT protocols which are categorized based on their application and the OSI model. 6LowPAN, IPv4, and RPL are some infrastructure-based IoT protocols.

(d) Cloud computing is a technology that dynamically performs end-users tasks using the concept of *virtualization*.

(e) Cloud is categorized based on the type of services provided by the cloud and the deployment.

(f) Mobile Cloud Computing (MCC) is another emerging technology which finds application in mobile devices. MCC improves storage capability, processing, and saves energy.

(g) Fog computing provides a distributed framework for distributing heterogeneous resources, reducing latency, and improving the processing of data.

(h) Sensor cloud provides different applications to multiple users at the same time using the concept of virtualization.

(i) Big data is the huge volume of data generated from the millions of connected IoT devices in a structured, semistructured, and unstructured format.

Exercises

1. Discuss the layered architecture of IoT.

2. What are the different application-based IoT protocols? Discuss their applications.

3. Which protocol establishes full-duplex communication?

4. What are the primary characteristics of the AMQP protocol?

5. What is IPv6? Discuss the features of IPv6.

6. What is cloud computing? Discuss the workflow among the various actors in the cloud infrastructure.

7. Write the differences between AMQP and WebSocket protocol.

8. How are the different types of clouds categorized based on cloud computing services? Discuss.

9. Discuss the benefits of the business perspective of the cloud.

10. What is virtualization? What are the advantages of virtualization?

11. What is MCC?

12. Write the differences between MCC and cloud computing.

13. Write the differences between fog computing and cloud computing.

14. What is the sensor cloud? Discuss the working principle of the sensor cloud.

15. What is big data? Discuss the characteristics of big data.

16. Write the different actors of the sensor cloud and briefly discuss them.

Introduction

Introduction

2

Overview of Industry 4.0 and Industrial Internet of Things

Learning Outcomes

■ This chapter covers the basic features of the Industry 4.0 and Industrial Internet of Things (IIoT).

■ New readers will be able to have a glimpse of the history of the industrial revolution and the evolution of Industry 4.0.

■ Experienced learners will be able to touch-up the various stages of the industrial revolution.

■ The basic concepts of the Industrial Internet, Cyber-Physical system, and the interlink of IoT, CPS, and IIoT is discussed in this chapter.

■ The key application areas of IIoT and Industry 4.0 are also discussed in this chapter.

2.1 Introduction

The process of transformation in technologies, which brings significant changes in the economic, cultural, and social structures of humans, is known as *Industrial Revolution*. The advancement in technologies over the last 200 years is responsible for the metamorphosis from traditional industrial systems and has steered the evolution of smart manufacturing systems, smart factories, smart healthcare, smart agricultural practices, and autonomous cars. Various technologies, such as Artificial Intelligence (AI), robotics, Internet of Things (IoT), 3D printing, nanotechnology, and Quantum computing, led to the development of many advanced systems. These gradual improvements in technologies have resulted in the 4th *Industrial Revolution*, also termed as *Industry 4.0*. For example, in the generation of Industry 4.0, standalone systems such as cranes, vessels, automated guided vehicles (AGV), sensor nodes, raw materials, and products will communicate among them in a port enabled with advanced technologies. A central system will help the logistics and other devices to coordinate and communicate among them to reduce the waiting time for container loading and unloading. Therefore, the overall incorporation of the different advanced technologies in the field of computation, networking, and physical processes, with the integration of Internet connectivity among the processes, promotes the development of *Cyber-Physical Systems (CPS)*. These systems operate and interact with the physical world in real-time. The components of CPS are developed by human beings, which include processing, communication, and information control. As described by Accenture, Industry 4.0 is the abbreviated name for the digitization in the industries [37]. Further, to cope with the rapid changes throughout the world, the digital transformation of the sectors can be a solution. This digitization is represented as the present state of the art and termed as Industry X.0. As per our opinion, the wave of technological innovations has brought widespread changes in the industries. On the other hand, the connectivity and interoperability among heterogeneous devices enables Machine-to-Machine communication (M2M), which enhances the safety of machines and workers. The smart machines are capable of detecting any fault in the system and accordingly apply corrective actions. Due to M2M communication, most of the hazardous works are performed by machines. Therefore, the health hazards caused to the workers are significantly reduced. Additionally, this minimizes the downtime of machines and improvement in the overall efficiency and system performance.

The integration of the smart sensor nodes, software, communication and networking technologies, cloud service, and storage systems forms the *Industrial Internet*. General Electric (GE), one of the founding members of *Industrial Internet Consortium* (IIC), coined the term *Industrial Internet* in late 2012. The three key factors of the Industrial Internet are: intelligent and interconnected machines, complex analysis of the huge amount of data generated, and connected workforce. The global market of Industrial Internet of Things (IIoT) is presumed to annually increase by 8.06% from the year 2018–2023, across the various industries such as manufacturing, oil & gas, healthcare, transportation, logistics, and mining. Further, the emergence of real-time supply chain improves the IIoT market through tracking of market, materials, and equipment. However, the growth of the IIoT market is still limited by the security-related issues associated with Big Data and AI [38].

2.2   Industry 4.0

Industry 4.0 is the term assigned to the present technological trends such as automation, exchange of data among the interconnected devices, and interoperability. Industry 4.0 is also referred to as the 4th Industrial Revolution. Further, it incorporates various enabling technologies into the industrial manufacturing processes such as CPS, Cloud computing, and IoT. Interoperability among various processes, devices, collected data, and real-time processing of the data generated from the sensor nodes are the key factors addressed during the evolution of Industry 4.0. Fig. 2.1 illustrates the combination of the heterogeneous machines, workforce (humans), and systems in the industries. Industry 4.0 introduces Internet-based technologies to improve communication among the systems. The communication among the interconnected devices, systems, and machines collectively leads to the formation of Industry 4.0.

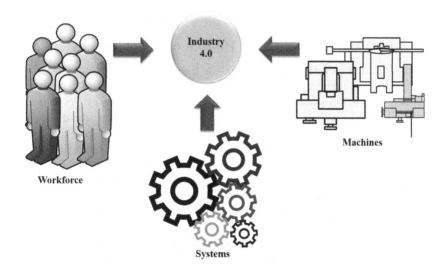

FIGURE 2.1: Industry 4.0.

The essential design characteristics of Industry 4.0 are as follows:

(a) Heterogeneous devices such as sensor nodes, actuators, devices, and machine are interconnected and provided a unique identifier via Internet-based technologies. Further, these interconnected devices allow interoperability to form the Industry 4.0-enabled production line.

(b) The information system aggregates the raw sensor data and creates the virtual copy of the physical plant using the data collected.

(c) The interconnected systems can communicate among themselves, execute complex tasks in real-time, and make decisions with very little or no human invention. These jobs performed by machines may be unsafe for humans.

(d) The systems support humans by collecting, processing, and analyzing the data received from the sensor nodes to generate decisions on their own.

From a historical point of view, the industrial revolution has transferred the technological shift of human beings from the domestication of animals to the development of technologies in the industry. Although the various industrial revolutions started years back, the effects of the revolution on the economy, social structures, education, medicine, and culture were long-lasting and significant. It is also characterized by the inception of new business models, reference architectures, modification of the manufacturing, supply chain, and transportation systems.

2.2.1   Industrial revolution: Phases of development

The Industrial Revolution has resulted in the all-round development of economic and social structures in different phases. As shown in Fig. 2.2, the first industrial revolution started in the 18th century. In the early 21st century, the fourth industrial revolution was officially initiated. With the advancement in industrial developments, the stages of development of machines during the first generation to the cyber-physical based systems during the fourth generation is shown in Fig. 2.3. The evolution of the first industrial revolution occurred in Great Britain, where the entrepreneurs were the pioneers of this sudden technological development. After that, mass production played a key role during the second industrial revolution. The advent of the third industrial revolution led to the evolution of wireless communication technologies (1G). Further, around 20th century, smartphones came into existence with in-built video calling facilities. The industrial revolutions also resulted in the transfer of knowledge among the workers associated with these innovations.

First Industrial Revolution: The first transformation in human civilization occurred from seeking food to farming practices, which was initiated around 10,000 years ago. The initial step was the domestication of animals for farming, production, and transportation. From 1750 to 1840, new technologies emerged in the manufacturing process. The first industrial revolution marked the conversion of muscle power to machines, as depicted in Fig. 2.3. Moreover, the efficiency of water power generation also improved during this period. New industrial manufacturing and iron production processes were developed in this period. The first industrial revolution paved the path toward technological developments such as mechanization of cotton spinning with the help of steam, improvement in the efficiency of steam engines, and reduction in the fuel cost of iron and wrought iron production. Charles Babbage, known as the father of the computer, invented the first mechanical computer, during the period of 1822–1834.

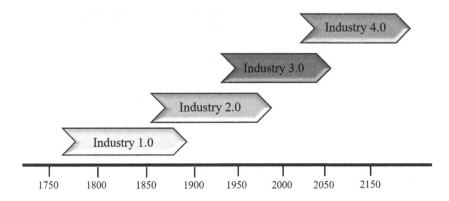

FIGURE 2.2: Development phases of industrial revolution.

Second Industrial Revolution: The two popular energy forms used throughout the second industrial revolution were petrol and electricity. Another critical and remarkable technological trend developed during the second revolution was mass production. Being inspired by the pulley system of the slaughterhouses in the United States, Henry Ford started the Ford Motor Company. Later, the assembly line was incorporated in the manufacturing industry, which reduced the number of workers required to complete any particular job. Moreover, funds were also raised for social welfare, which improved the health conditions of the common masses. During the later stages, from 1980 to 1990, the first wireless communication technology (1G) evolved. The cell phones developed were big, with poor voice qualities and battery life.

Third Industrial Revolution: With the beginning of the 20*th* century, the third industrial revolution started. Automation in the manufacturing system, computers, the Internet, and usage of renewable energies are some of the advancement in technologies, which took place during the third industrial revolution. Therefore, this revolution is also known as the Computer or Digital Revolution. After the second industrial revolution, the development of vacuum tube-based computers was initiated. Further, smart computers with integrated circuits and microprocessors emerged during this period. Smartphones also evolved in the market, armed with 3G facilities. 3G or 3rd generation is defined with the increase in bandwidth, faster communication, and video calling facilities. Business facilities also improved with the initiation

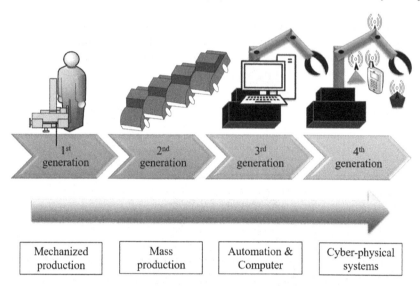

FIGURE 2.3: Industrial revolution.

of insourcing* and outsourcing.† Hence, the process termed as *globalization* was initiated. Globalization interconnects trade and cultural activities and improves the overall production of goods and services around the world.

2.2.2   Evolution of Industry 4.0

The extensive use of digitization and Internet-based technologies are the key characteristic features of *Industry 4.0*. The 4th industrial revolution is termed as Industry 4.0 [41]. The various features of Industry 4.0 are as follows:

(a) Integration of various technologies such as AI, Quantum computing, gene sequencing, augmented and virtual reality, 5th generation (5G) wireless communication technologies, and 3D printing formed the stepping-stone toward the evolution of *Industry 4.0*. These technologies form the building blocks for smart factories, smart and connected machines, and smart manufacturing systems.

(b) The basic skeleton of Internet-based technologies is smart sensor nodes. The smart sensor nodes evolved to be powerful, cheaper, and smaller in size, compared to the traditional sensor nodes.

---

*Business practices within the same country, which employ persons for other company.
†The business processes or services, which are carried outside the country for the reduction in costs.

The integration of IoT and Industry 4.0 assisted the emergence of the IIoT, as depicted in Fig. 2.4. The interconnected devices, machines, and things are identified by a unique identifier and combined with automated machines. This connected network of things forms the basic foundation of IIoT. In other words, the applications of IoT in industries paved the way for the advent of IIoT. On the other hand, 5G wireless communication technologies act as a key feature for both Industry 4.0 and IIoT. These 5G technologies are characterized by high speed, faster data transmission, assist in interactive multimedia, and voice streaming. The overall technological changes will have a greater impact on the entire business ecology. Further, the integration of these advanced technologies with the existing automation systems will initiate massive digitization in industries. The dissimilarity between traditional automation and Industry 4.0 is given in Table 2.1. On the other hand, Industry 4.0 and IoT applications are redefining automation in the industrial sectors. Further, the concepts of a digital twin of the products and processes emerged with Industry 4.0. The terms IIoT and Industry 4.0 are mostly discussed together and interchangeably in the technological sectors. However, there are certain differences between Industry 4.0 and IIoT, which are listed in a tabular format in Table 2.2.

Professor Erik Brynjolfsson (Director) and Andrew McAfee (Principal Research Scientist), Massachusetts Institute of Technology (MIT), Center for Digital Business, coined the term *Second Machine Age* for the recent advances in computational power, communication technologies, and data storage [39]. The *intelligent machines* of this age are capable of performing various tasks that humans can do, thereby reducing the number of workforces required on the factory floor. Therefore, the 4th Industrial Revolution can be termed as the Second Machine Age. New durable and lighter materials are being developed and produced during this revolution. The various features of this age are processing, storage and analysis of Big Data, M2M communication in the industries, the safety of machines and workers, connectivity among the devices and things, and data security. These diverse features of Industry 4.0 are shown in Fig. 2.8(a).

2.2.3 Environmental impacts of industrial revolution

The industrial revolution has brought remarkable changes in technologies and energy usage. On the other hand, the after-effects of this industrial revolution have resulted in increased global pollution levels, resulting in significant environmental degradation. Further, with the emergence of machines, air pollution has increased. The steam engines and various industrial processes consume coal, which emits an enormous amount of fumes and other pollutants. These raise the air pollution level to a greater extent, specifically in the industrial regions. Additionally, the highly toxic chemical outcomes from different industrial processes are dumped into the water bodies, which further results in water pollution. These toxic gases and effluent discharges trigger

Table 2.1: Comparison between Industry 4.0 and traditional automation [44].

Features	Industry 4.0	Traditional Automation
Real-time communication	Machine-to-Machine (M2M) communication is possible with minimum latency (dynamic)	Machine-to-Machine (M2M) communication occur in small scale (mostly static)
Digital information	One of the actor, either sender or receiver is a machine	Sender or receiver may be a machine or any device
Automation	Automation is possible among the machine parts with the help of logic system	Automatic operation of machine parts is not possible

Table 2.2: Comparison between Industry 4.0 and IIoT [45].

Features	Industry 4.0	IIoT
Initiative	German government initiative (also known as Fourth Industrial Revolution)	GE termed it as Industrial Internet, while CISCO termed it as Internet of Everything. The IIC recognizes IIoT.
Primary focus	Manufacturing Industry	Both manufacturing and nonmanufacturing industries
Objective	To secure a competitive position in a dynamic market	To enable and accelerate the Internet-enabled technologies across industries
Description	Economic impact	Technological development
Effect	Less advantageous in the long run	Advantageous in the long-run

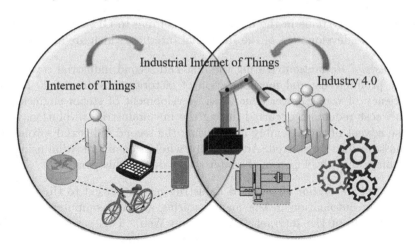

FIGURE 2.4: IoT and Industry 4.0.

environmental degradation. The usage of fossil fuels has also increased with the advent of the industrial revolution. As a result, there is a considerable depletion of globally limited natural resources. Another major impact of industrialization is global warming. The smoke produced from the industries in combination with the other harmful greenhouse gases released into the atmosphere gives rise to global warming.

2.2.4 Industrial Internet

The term *Industrial Internet* was coined by the U.S. corporation, GE. Industrial Internet resulted from the combination of physical and digital worlds. As defined by GE, Industrial Internet is "the convergence of the global industrial systems with the power of advanced computing, analytics, low-cost sensing, and new levels of connectivity permitted by the Internet." The basic concept of the Industrial Internet is illustrated in Fig. 2.5. The Industrial Internet-enabled organizations use sensor nodes, software, and Machine-to-Machine (M2M) communication to collect data from material things or devices. As shown in Fig. 2.5, the interconnected devices in industries, automated assembly line, real-time data collected from sensor nodes, and their analysis together can be referred to as the Industrial Internet. The processed and analyzed data provide value-added services [42]. Industrial Internet will lead to an increase in the speed and efficiency in a wide range of industries such as aviation, railways, mining, power generation, and healthcare systems. Additionally, it will result in improved job facilities, accelerate the productivity of the manufacturing system, and promote all-round economic growth. In simple words, the industries will have optimized and centralized management of the equipment, processes, and systems in the industries. According to GE, there exist three phases in the development of the Industrial Internet, which are as follows:

(a) *Industrial Revolution*: During the first and second industrial revolution, the productivity and mechanization of factories were initiated. The efficiency of water power generation, development of steam engines, and fuel cost reduction occurred during the first industrial revolution. With the advent of the assembly lines during the second industrial revolution, mass production started. Additionally, wireless communication using cell phones were developed in the later stages during this period.

(b) *Internet Revolution*: The third industrial revolution led to the development of automation in the manufacturing system, computers, and the emergence of the internet (World Wide Web). This revolution is also known as the Computer or Digital Revolution. One of the most remarkable effects of this revolution was information exchange over large distances through the internet and smartphones.

(c) *Industrial Internet*: The amalgamation of CPS and the evolution of Internet-based technologies led to the development of the Industrial Internet. Therefore, it is generally assumed that Industry 4.0 or the 4th industrial revolution evolved from these technologies.

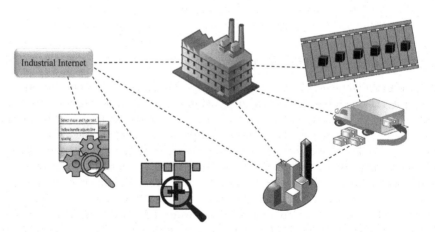

FIGURE 2.5: Industrial Internet.

Intelligent machines,[‡] advanced analytics, and connected people[§] are the key elements behind the development of *Industrial Internet*. The interconnection among these key elements and their combination improve the overall economy through optimization of the processes, minimization of product costs, and real-time analysis of data. The real-time data helps to provide deeper insights and improve the system operation in various industrial sectors. The catalysts behind the development of *Industrial Internet* are as follows:

(a) The investment in the deployment of sensor nodes, required instruments, business processes, and updated user application interfaces will promote the development of new technologies.

(b) The security of sensitive information, data, and intellectual property (IP) requires a robust security system.

(c) To instigate the growth of new talents such as advanced engineering studies,[¶] data scientists, and software experts promote industrial development.

2.2.4.1 Applications of Industrial Internet

Industrial Internet has brought a wave of transformation in the industrial sector with the help of available resources such as smart sensor nodes, software, and M2M communication [42]. It has a wide range of applications in various industries which are as follows:

Aviation industry: Industrial Internet helps in the management of crews, reduction in fuel consumption, maintenance of the flight engine, timely flight repair, and cancellation and schedule of the flights. The optimization of operation and assets upgrade the safety of workers at each of the stages of operation.

Transportation: The real-time analysis of collected data, predictive analysis of the data, and software availability result in the reduction of the maintenance costs, breakdown of engine, and optimization of train schedules. The optimal scheduling of trains results in cost optimization, optimum supply management, and maintenance of timing. Further, with the integration of advanced technologies, the on-road safety of vehicles and drivers can be improved.

Healthcare: With the exponential growth in the world's population, it is quite challenging to maintain healthcare standards. Internet connectivity has resulted in the availability of the reputation of doctors, upgradation of the

---

[‡]The integration of smart sensor nodes, control systems, and various software applications leads to the formation of intelligent machines.

[§]The term *connected people* denotes the monitoring of interconnected common people at the office, workplace, and home using sensor nodes to enhance their safety and provide higher quality services.

[¶]The combined study of mechanical and industrial engineering, which provides students a detailed knowledge of both the sectors.

healthcare system, and improvement in the treatment outcomes. Thus, the Industrial Internet enables a safe, secure, and efficient healthcare system.

Power Production: To enhance the production of power, maintain reliability, safety, and optimize the utilization of fuel are the primary objectives of the Industrial Internet. Power outage is a common problem throughout the developing world. To detect the location of the faults in power lines, or the faulty transmission line equipment is a difficult task. However, with the help of the Industrial Internet, as each of these devices is interconnected, the detection of faults can be managed in real-time.

2.2.5   Applications of Industry 4.0

Industry 4.0 is the integration of the network of humans, things, machines, devices, and robots, which interact with each other and share information among them. This network digitizes the manufacturing system and analyzes the real-time data available with the CPS [43]. The efficiency of the product improves in terms of time and cost, quality, and overall productivity enhances. Further, interoperability among the things is one of the main features of Industry 4.0. The various applications of Industry 4.0 are as follows:

Smart Factory and Manufacturing system: With the evolution of advanced technologies such as AI and machine learning, competitiveness in the market, risks associated with failure, and increase in production assist in the development of the smart factory system. Smart devices, smart engineering, information technology, complex data analysis, and innovation are the different components of a smart factory. Smart factories and manufacturing systems reduce the manufacturing cost, increase overall efficiency, enhance safety, and improve the quality of products.

Smart Product: Industry 4.0 comprises various technologies such as CPS, Information and Communication Technologies (ICT), IoT, and cloud-based data integration to help human beings and things to communicate among themselves. These advanced technologies, along with smart sensor nodes, form the *smart product*. Further, it includes advanced knowledge, information, and real-time production system. Smart products are fabricated using interlinked physical and digital processes.

Smart City: Smart people, smart mobility, smart environment, Internet, sensor networks, wireless broadband network, and smart governance converge together to form a *Smart City*. The primary goal of a smart city is to improve the quality of life for everyday people, the safety of citizens, better and timely transportation, and increased energy efficiency. However, the transformation of a traditional city into a smart city is a time-taking and complicated task.

Lean Production system: The lean production system is a process of waste minimization in a manufacturing system while maintaining the quality of the product. There exist seven types of wastes associated with the lean manufacturing system – excess inventory, movement of people or information, waiting time, non-value-added people, over-processing, defects in products, and overproduction. Given Industry 4.0, a lean production system identifies the wastes, continuously improves the product quality, reduces the cost of the product, and adds value-added activities. Further, it also minimizes the lead time.‖

Smart and Connected Business: Smart and connected products enhance efficiency and overall production. With the incorporation of a smart business model - processes become cheaper, the efficiency of the process increases, and fulfill the expected revenue. Smart and Connected products are characterized as monitor, control, optimization, and automation. The critical properties of smart and connected business are business statements that add value to the product, revenue streams of a company or organization, and technologies.

2.3 IIoT

IIoT is the application of different IoT-based technologies, such as M2M communication, AI, Machine Learning, Big Data technology, and various automation technologies, in the industrial sector. The primary focus of IIoT is to improve the manufacturing and production processes. IoT and IIoT devices, automation levels, and application purposes are different. The various key differences between IoT and IIoT are discussed in Table 2.3. IIoT can also be referred to as a larger subset of IoT. Fig. 2.6 shows the basic workflow in an IIoT-enabled industry. The key enablers of IIoT are smart sensor nodes, smart machines,** automation, optimization, and real-time monitoring, which create industrial transformation. As shown in Fig. 2.6, the sensor nodes deployed at various industrial locations sense and transmit data to the server. The real-time processing and complex analysis of the data help to optimize various industrial processes such as predictive maintenance of machines, inventory management, and packaging of finished products. With the popularity of IIoT technologies, smart sensor nodes and processors have become available and affordable. Further, IIoT is not an interconnection of physical devices, rather than the digitization of products, processes, and factories. The IIC and GE collectively conclude that IIoT is a synonym for the Industrial Internet. IIoT utilizes the principles of CPS to detect the overall processes of the plant and take automated decisions based on the real-time data obtained. The primary goal of IIoT is to improve operational efficiency and optimize the costs, thereby re-modifying the reference architecture and business models. The benefits of IIoT drivers are as follows:

(a) Operational efficiency of the different industrial processes is improved.

‖Lead time is the time-delay between the initiation and execution of the process
**Machines which are capable of communicating among themselves.

(b) Product development and assembly line-related costs (operational cost) are reduced.

(c) Downtime[††] of machines is reduced, and energy is conserved. In addition to this, machines can schedule their maintenance using predictive and prescriptive analysis of the data.

(d) The managers and other relevant officials can remotely monitor the factory operations.

(e) Improved estimation of available materials, work progress, and the arrival of inventory lead to the maximum utilization of the resources.

(f) Safety and security of workers is enhanced.

(g) Both customers' experience and cost of packaging the products have improved significantly.

(h) New business opportunities have been created.

(i) The real-time monitoring of the supply chain reduces inventory and capital requirements.

Therefore, IIoT is typically applicable for industrial purposes. IIoT is widely applicable in the energy, military, safety, healthcare, and transportation, where reliability and accuracy are essential aspects to be maintained. There are some existing traditional automation systems, which can keep track of the overall production process inside a plant. However, IIoT additionally provides the intricate details of the activities.

2.3.1   Prerequisites of IIoT

IIoT can be described as an interconnection of a large number of industrial processes and systems, which communicate and coordinate among themselves. The real-time data collected from sensor nodes are stored, processed, and analyzed to improve the performance and efficiency of the overall system [46]. As the digital world and the physical world can communicate among them through sensor nodes and actuators, industrial systems have obtained a close resemblance to the CPS. Further, the latency in the IIoT networks is maintained in less than milliseconds, which otherwise may possess a threat toward the health and safety of machines and workers. The various requirements of IIoT [47] are as follows:

(a) Security: Security is one of the important aspects of IIoT. Any disturbance in the production process may lead to a loss of millions of dollars. For

---

[††]Time for which the machine was unavailable or not in operation.

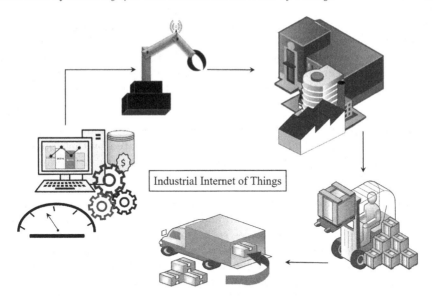

FIGURE 2.6: Industrial internet of things.

example, any failure in the smart grid creates a blackout over a large region in a country. As a consequence of the blackout, millions of people are affected, and national security is endangered.

(b) Interoperability: In an industrial system, various technologies such as Supervisory Control and Data Acquisition (SCADA) and M2M coexist together to create an advanced manufacturing system. Moreover, semantic interoperability allows the exchange of data between machines in an unambiguous format. On the other hand, semantic interoperability requires syntactic interoperability to transmit data. For example, the transmission of healthcare data in HL-7 format is performed to maintain data uniformity and interoperability.

(c) Scalability: Heterogeneous machines, robots, and controllers are connected in large numbers to form an industrial network. Therefore, almost thousands of heterogeneous sensors in a network collect, analyze, and processes data continuously at the same time. For an end-to-end solution, synchronization among the components of the system is necessary.

(d) Accuracy: In industries, a huge volume of data is generated every millisecond from connected devices, machines, and processes. Proper synchronization of the collected data is essential to maintain product quality. Based on the variations in data, immediate corrective measures need to be taken to improve the product quality and system efficiency.

(e) Latency: In any industrial process, for each of the operational seconds, the sensor nodes sense and transmit the data. In case of any observed anomaly, the data should be corrected through automated processing. If any delay occurs in the detection of data, processing, and execution, it will compromise the safety of workers, product quality, and costs.

(f) Reliability: In industries, the systems operate under severe operating conditions such as extreme heat, vibrations, pressure, dust, and cold, for long periods. These systems are mostly remotely operated by their operators. Therefore, these systems should have high uptime, greater tolerance, and reliable operation for hours.

(g) Automation: Most of the industrial processes are automated from beginning to the end of the operation, and require minimum or no human intervention. Certain technologies such as deep learning, automation logic, control processes, and various data analytics methods help in incorporating intelligence into the edge devices. Further, new systems should be easily programmed and integrated with the existing systems.

(h) Serviceability: The sensor nodes, actuators, and the protocols form the base of any industrial system. Upgradation of the firmware is necessary to configure the gateways and servers, to maintain performance levels, and meet the maintenance requirements of the system.

2.3.2 Basics of CPS

CPS integrates physical objects and computer-based algorithms, which communicate among themselves and acquire knowledge from their surroundings. The interconnection among the different objects such as sensor-enabled machines, processes, storage, people, and the network edge is illustrated in Fig. 2.7. These physical objects and the users are connected with the help of the Internet. For example, the Intel processor developed in the year 1971 had 2300 transistors working at a frequency of 740 Hz. On the other hand, the current Intel processors have more than 2 billion processors, which operate at a frequency of about 3 GHz. CPS combines and coordinates among the physical and digital elements [48]. The characteristics of the interconnected elements are categorized as follows:

(a) Compute capabilities: Higher processing power, cheaper, and smaller in size.

(b) Sensing technologies: Sense and transmit data from the environment in real-time.

(c) Connectivity among the elements: The elements seamlessly connect with the multiple technologies at the same time instant and apply feedback control.

Table 2.3: Comparison between IoT and IIoT.

Features	IoT	IIoT
Focus	Devices are developed for the well-being of individuals	Devices are developed for the improvement of efficiency, safety, and security in industries
Interoperability	M-2-M communication possible in small scale	M2M communication, SCADA, and other manufacturing technologies are applicable in large scale
Scalability	Applicable in small scale, compared to IIoT, where things are interconnected through the Internet	Applicable in large scale networks, where seamless integration of devices are done in the industrial networks and IoT devices
Automation	IoT devices sense, process, and transmit data with limited or no human involvement	IIoT devices, in addition to the intelligence of the IoT devices, require control and automation logic, deep learning, and analytics
Emergency situation	In case any device fails, an emergency situation will not occur	Failure of any device may lead to a critical situation
Application	Consumer purpose	Industrial purpose

Traditional CPS mostly depends on a network of interconnected objects, which are static. A CPS system possesses certain basic characteristics such as sequential input–output, communication among multiple processes, real-time feedback control based on the data collected from sensor nodes attached to the machines, synchronous and asynchronous operation, and applicable to safety-critical applications. Fig. 2.8(b) briefly describes the different characteristics of CPS. However, with the mobility added to the physical devices, the systems are ubiquitous and integrated seamlessly.

Another form of CPS is *Mobile Cyber-Physical systems*, where the physical system is mobile. The computational resources, communication radio, sensor modules, and high-level programming facilitate mobile devices to be categorized as mobile CPS. Mobile CPS has gained much popularity due to the following features:

(a) Several input-output devices are interconnected, which communicate among themselves with the help of Internet.

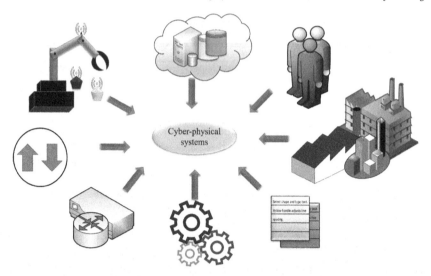

FIGURE 2.7: Cyber-physical systems.

(b) Presence of significant processing devices such as local storage in a smart-phone and feedback-based control systems.

(c) Programming languages and automation logic encourage faster development of mobile CPS.

The primary constraints of mobile CPS are the stability of the network, power requirement, security, and dynamic nature of the system. Mobile CPS has a wide range of applications, which are as follows:

(a) Mobile robots: multiple smart sensor nodes, mobile devices, and complex analytics help to enhance the efficiency and process of complex tasks.

(b) Intelligent transportation system: communication among the vehicles and with the infrastructure improves the safety and security of the vehicles.

(c) Smart technologies: smart city, smart healthcare system, and monitoring of environment help to upgrade the safety, security, and reduce energy consumption.

2.3.3 CPS and IIoT

CPS integrates physical objects with digital objects through the Internet. The digital objects virtually act as the physical machine. The network of these connected objects in the manufacturing industries is implied as to the IIoT. Therefore, the connected CPS, which is applicable in the industrial sector, is

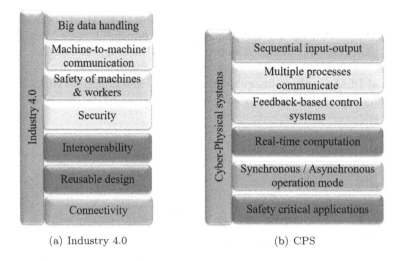

(a) Industry 4.0   (b) CPS

FIGURE 2.8: Characteristics of Industry 4.0 and CPS.

termed as IIoT. As shown in Fig. 2.4, the convergence of IoT and Industry 4.0 leads to the development of IIoT. Fig. 2.8 shows the characteristics of CPS and Industry 4.0. The machine is first connected to the local server, where the sensed data is transmitted. The sensed data is then converted to machine-level data. The adaptive health of the machine is assessed through the digitally virtual object created. Machine quality, condition, and quality of the product are checked. Finally, the machine is configured based on the processed data.

IoT is the necessary infrastructure for Industry 4.0 or Fourth Industrial Revolution, CPS, and IIoT. IoT enables the remote sensing of physical objects across the whole network. Further, CPS is the foundation block to build future smart factories. Various forms of CPS-based applications are utilized to develop Industry 4.0. The smart machines are self-configuring, self-optimizing, and can adjust themselves with the sudden fluctuations in the network. On the other hand, smart analytics, deep learning, and machine learning methods provide real-time information about the machine's health and any form of degradation in the machine condition. Therefore, industrial CPS boosts the production process, reduces downtime, decreases machine set-up time, minimizes errors, and garners faster responses. The quality of the product improves through continuous improvement using feedback-based control and advanced analytics.

2.3.4 Applications of IIoT

IIoT has a wide range of applications across various manufacturing industries. The application of Computer-Integrated Manufacturing (CIM) systems on the shop floor of industries has led to the development of Product Data Management (PDM).* PDM forms the pillar for the Product Lifecycle Management (PLM).[†] On the other hand, in industries, PLM acts as the interface for the different applications during the entire lifecycle management of the product. PDM and PLM are the primary requirements for the growth of IIoT. The key application areas of IIoT are as follows:

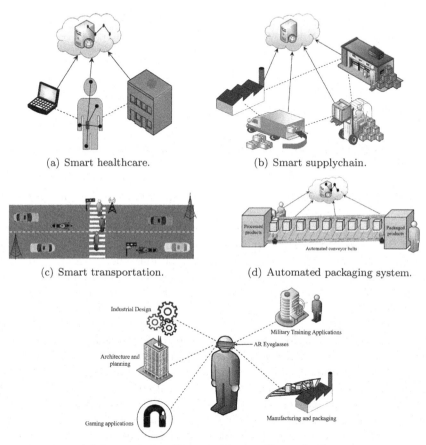

(a) Smart healthcare.

(b) Smart supplychain.

(c) Smart transportation.

(d) Automated packaging system.

(e) Augmented reality and virtual reality applications.

FIGURE 2.9: Applications of IIoT.

---

*PDM is a technique applied for designing interconnected networks among the workers, data, and machines.

[†]PLM is the process of management of the sequential steps for the development of a product from its specification and design, to delivery of the product.

(a) Smart healthcare: The interconnected network of devices and systems are dependent on newly emerged smart sensor nodes, medical devices, and systems. The health conditions of a critical patient may fluctuate at any time instant. Therefore, continuous monitoring of the patients is essential. The different types of intelligent devices are dynamically configured as per the patient's requirement. Augmented and virtual reality devices assist the doctors during surgery and therapy. The major challenge in the further advancement of these devices is to design and develop a safe, secure, and reliable healthcare system. Fig. 2.9(a) illustrates a healthcare system, where the sensor nodes sense and transmit the physiological data of the patient to the local processing unit (LPU). Further, the LPU transmits the data to the local server. On the other hand, medical experts such as doctors, paramedics, and nurses in hospitals can remotely observe the health conditions of the patient.

(b) Smart supply chain: Intelligent, interconnected, and real-time information of the product has encouraged the emergence of the smart supply chain.* The requirement of proper maintenance of the raw materials, available inventory stock, details of each steps involved in the production process, proper flow of information among the various stages, maintaining the time-window for delivery of goods, and returning the faulty goods are the essential elements involved in the development of a smart supply chain. Fig. 2.9(b) illustrates the proper flow of information among the various components of the supply chain. The real-time data handling, processing of data, and storage are done centrally at the cloud. Further, automation is involved in the various processes of the warehouse such as inventory management, loading of products, and distribution.

(c) Smart Transportation: With the rapid increase in the number of on-road vehicles, the safety of the drivers and vehicles has become an issue of major concern in the road transportation industry. Intelligent Transportation System (ITS) and Advanced Driver Assistance Systems (ADAS) have been developed to improve the traffic system. Further, these technologies have led to a reduction in on-road congestion, accidents, and a reduction in the casualty rate. Fig. 2.9(c) portrays a smart and connected transportation scenario. The sensor nodes are placed on the vehicles and Road Side Units (RSU). These deployed sensor nodes sense and transmit data to the local server. Various real-time information such as safe speed, safe distance with the neighboring vehicles, and weather conditions are conveyed to the drivers. Further, the communication among the vehicles and the infrastructure (RSUs) results in the improvement of the safety of vehicles, pedestrians, and drivers.

---

*Supply chain denotes the entire process involved in the development of a product or service, starting with the producer to the consumer.

(d) Smart manufacturing system: In order to improve the efficiency of production and product quality, reduce the per-unit cost of production, and enhance the lifetime of machines and developed products, the smart manufacturing system is developed. Fig. 2.9(d) depicts the automatic packaging of a product in a manufacturing plant with minimum human intervention. Different sensor nodes are deployed at various locations, and robots or machines pack products. As we can observe, the conveyor belts adjust and deliver the products as per the requirement. On the other hand, the operators can remotely monitor the entire process. Therefore, smart sensor nodes, actuators, and various software-based algorithms form the base of the manufacturing system. The integration of these technologies and interconnected physical devices allows simulation, optimization, and communication among the various processes during the operational period of the plant. The network among these devices and virtual models forms the smart manufacturing system.

(e) AR and VR applications: Augmented Reality (AR)[†] and Virtual Reality (VR)[‡] have widespread applications in the optimization stages of manufacturing industries, inventory management in warehouses, training of personnel in military, healthcare, and assembly line operations [49]. AR makes use of the existing environment and covers the information above while VR allows the end-user to be emerged into the environment, with the help of auditory and visual feedback. Fig. 2.9(e) shows the applications of AR and VR in different fields such as architecture design, planning, manufacturing and packaging, military training, and entertainment, using smart eyeglasses.

(f) Smart Grid: Automation and usage of renewable energy started with the third industrial revolution. However, with the advent of digitization, the energy efficiency and power generation have improved. Through the scheduling of devices, demand optimization and distributed automation have resulted in a significant reduction in the peak load, faults, and power shortages. The improvement in the Flexible AC Transmission Devices (FACTS) and Phasor Measurement Units (PMU) has produced improved solutions for the smart grid.

Check yourself

Basics of power transmission, advantages of A.C. transmission, supply chain management, AR, VR, PLM, and PDM.

[†]AR is an extended version of reality. It provides an augmented view of the physical world, with the help of computer-generated images.

[‡]VR provides a three-dimensional view of the object. An interactive and amalgamated form of the software and hardware-based artificial environment is presented to the user.

Summary

This chapter consists of the basic concepts of the Fourth Industrial Revolution or Industry 4.0, Industrial Internet, CPS, and IIoT.

(a) Industrial Revolution is the process of technology conversion for the changes in the economic, cultural, and social structure of human beings.

(b) With the first industrial revolution, the process of mechanized production evolved.

(c) During the second industrial revolution, mass production started.

(d) Automation in the manufacturing system, computers, and the Internet started in the time of the third revolution.

(e) The combination of various technologies such as Quantum computing, AI, 5th generation wireless communication resulted in the development of the 4th industrial revolution, or Industry 4.0.

(f) Industry 4.0 has a wide range of applications in the field of smart factory and manufacturing systems, smart products, smart cities, and lean manufacturing systems.

(g) The term Industrial Internet was coined by GE, which highlights the importance of the integration of physical and digital worlds.

(h) Industrial Internet is widely applicable in aviation, transportation, healthcare, and power production.

(i) IIoT, the larger subset of IoT, primarily focuses on the improvement of the manufacturing and production process.

(j) CPS integrates the physical objects and the computer-based algorithms through Internet.

(k) IoT is the basic foundation block of CPS, Industry 4.0, and IIoT.

(l) IIoT is widely applicable in the various industrial sectors such as healthcare, transportation, manufacturing, and supply chain.

Exercises

1. What is an industrial revolution? What are the different phases of development in the industries?

2. Write the differences between Industry 4.0 and IIoT.

3. What are intelligent machines? In which phase of industrial revolution intelligent machines came into existence?

4. What is the industrial internet? What are the main catalysts behind the industrial internet?

5. Discuss the application of industrial internet in (a) aviation, (b) transportation, (c) healthcare, and (d) power industries.

6. What is a smart product?

7. What are the advantages of IIoT drivers?

8. What is a Cyber-Physical System (CPS)? What are the primary constituents of mobile CPS?

9. Write a real-life case study on the application of IIoT in smart transportation.

10. Write the differences between IoT and IIoT.

3

Industry 4.0: Basics

Learning Outcomes

■ This chapter covers the impact of industrial revolution, evolution, requirements, drivers of Industry 4.0.

■ New readers will be able to gain knowledge of the various aspects of Industry 4.0.

■ This chapter also gives an insight into the sustainability assessment in industries, smart business perspectives, and impacts of Industry 4.0.

■ Various constituents of cyber-security and the cyber-threats are discussed in this chapter.

3.1 Introduction

The major technological shift toward the digitization of the manufacturing process in the industries laid the stepping stone towards the evolution of Industry 4.0. The term "Industrie 4.0" emerged from a technical project of the German government, which promoted the concept of automation in the manufacturing industries. Industry 4.0 has renovated the traditional manufacturing technologies into a faster, automated, flexible, and efficient process of producing improved quality goods at cheaper rates. A significant leap in technology from mechanized production during the first generation to industrial automation was observed with the advent of the third generation communication technologies. The development of 5th generation ($5G$) communication technologies, AI, quantum computing, and other technologies, in addition to the previous advancement in technologies, formed the base for the technological progress in the manufacturing industries. The amalgamation of Cyber-Physical Systems (CPS), Internet of Things (IoT), and other technologies made headway toward the fourth industrial revolution. Therefore, the combination of the existing manufacturing technologies with the current innovations forms the base of the fourth industrial revolution. The fourth generation of the industrial revolution is also known as *Industry 4.0*. The interconnection among the sensors, data, and analytics acted as the catalyst for this industrial revolution.

3.1.1   Historical context

The transformation of technology was initiated with the mechanization in the production system during the first industrial revolution in the late of 18th century. The efficiency of steam engines and water power generation improved during this period. Further, with the advent of the second industrial revolution, mass production started in the early phase of the 20th century. During the third industrial revolution, automation, the Internet, electronics, and computers evolved, which is usually known as *digital revolution.* Additionally, semiconductors, microchips, and smartphones also evolved during this period. The application of the various technologies such as Machine-to-Machine (M2M) communication, AI, Machine Learning (ML), and analytics in the manufacturing and production industries led to the emergence of the fourth industrial revolution. The inclusion of these technologies with the traditional ones formed the base of technological trends such as smart factories, lean manufacturing systems, and smart supply chains.

   To improve the overall growth, such that the economy improves, either the cost of labor, productivity, or capital invested has to be increased. Thus, factory sites were historically associated with the location of the cheap availability of labor. The working conditions of the laborers, during the advent of industrialization, were quite dangerous, unhealthy, and monotonous. Most of the workers were unskilled and were not assured of job security and work safety. Even the children were a part of the workforce, often having to face ill-treated and working for long hours in hazardous environments. The wages of workers were quite low, compared to their working hours. Moreover, no compensation or medical expenses were provided to the workers in case of an accident or injury at the site. The life of these so-referred to as lower-class people was quite challenging during this period. Additionally, the industrial revolution had a lot of adverse effects on the environment. Water and air pollution levels rose at alarming rates, fossil fuels rapidly depleted, and global warming increased significantly. The growth of new cities was another impact of the industrial revolution.

3.1.2   Significant changes in the industry

Industrial Revolution has had a far-reaching impact on the various industries such as transportation, steel, textile manufacturing, glass making, mining, and agriculture. The significant changes in the various industries are shown

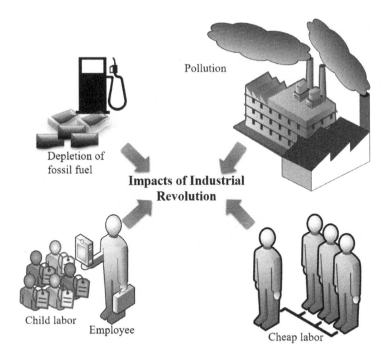

Pollution

Depletion of
fossil fuel

**Impacts of Industrial
Revolution**

Child labor

Employee

Cheap labor

FIGURE 3.1: Impact of industrial revolution.

in Fig. 3.2(b). With the advent of the metal production process during the end
of 18th century, iron, copper, and steel industries have had an important role
in the changes in the country's infrastructure – coal played a key part in the
industrialization of cities as well as countries. In the steam industries, coal was
widely utilized. Moreover, with the significant changes in the steam engine,
the transportation industry experienced notable variations. Before the development of steam engines, raw materials, finished products, and other goods
were transported by horse-driven wagons. As a direct result of the steam engines, transportation became cheaper, easier, quicker, and highly reliable. The
development of infrastructures such as roads and iron bridges, even in remote
places, was undertaken massively. As the mechanization started, materials
became cheaper. This resulted in increased demand compared to the supply.
Laborers were forced to work for extended hours. To prevent the exploitation of laborers, new laws were laid. In the textile manufacturing industry,
previously wool was hand-woven, which, after industrialization, was woven by
spinning wool and the loom. While agricultural practices witnessed the development of new tools, fertilizers, and harvesting techniques, which resulted in
improved crop production.

3.2   Design requirements of Industry 4.0

There are various requirements for transforming the traditional industries into Industry 4.0 compliant ones [53], as shown in Fig. 3.2(a). The necessary design requirements of Industry 4.0 are as follows:

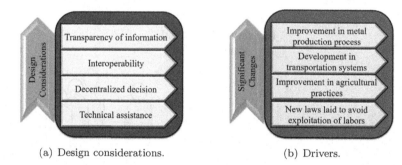

(a) Design considerations.   (b) Drivers.

FIGURE 3.2: Industry 4.0: Design considerations and changes brought in the industry.

(a) Transparency of information: The huge amount of data available in the industries provide information clarity to the operators and assist in data analysis. These data prove to be beneficial for the identification of gaps, innovation in methods, and improvement of the processes.

(b) Interoperability: The sensors, devices, people, and machines communicate with each other with the help of Internet-based technologies. The errors in communication due to misrepresentation of information and data are reduced due to interoperability.

(c) Decentralized decision: Smart machines are capable of taking self-decisions such that the tasks are independently performed by them through proper coordination among themselves (using M2M communication).

(d) Technical assistance: The assistance systems developed are competent enough to assist humans in their work and help to visualize the information. Further, these technically advanced systems also seamlessly perform certain tasks that are unsafe for humans.

3.3 Drivers of Industry 4.0

The industrial revolutions over the years have brought considerable technological changes to the industries. The inclusion of mechanization and automation into the industrial processes has resulted in the overall improvement of manufacturing industries. The various Internet-based advanced technologies has converted the industrial systems and processes into *smarter* ones. Technological advancements that have increased productivity are termed as the *drivers* of Industry 4.0. The classification of the drivers of Industry 4.0 is given in Fig. 3.3.

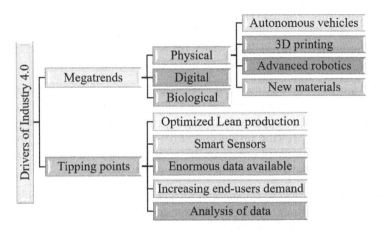

FIGURE 3.3: Drivers of Industry 4.0.

3.3.1 Megatrends

The format of technological advancements that have a relevant impact on the business models of the organization is termed as *megatrends*. There has been a noticeable leap in productivity after the initiation of the industrial revolution. Automation, increase in mechanization, digitization of the manufacturing industries, and the development of small-sized, cheap sensors formed the *technological trigger* behind the development of Industry 4.0 [51]. For example, the driver-less vehicles or autonomous vehicles are mostly dependent on AI, Deep Learning (DL), and smart sensor nodes. Further, CPS forms the building block of these megatrends, which allows for the integration of these physical devices with Internet-based technologies. Smart materials permit these sensors to utilize minimum energy and transmit data to the server.

Megatrends are classified as *physical, digital,* and *biological* trends. These three types of trends are interrelated among themselves.

Physical trend: The physical trend includes tactile technical innovations in the manufacturing industries. There are a few physical and technological advancements such as autonomous vehicles, 3D printing, advanced robotics, and new materials, whose development is easily associated with this trend.

(a) Autonomous vehicles: The development of smart sensors and the progress in AI applications have improved the development of automated trans-portation systems. Driverless vehicles such as trucks, cars, and drones efficiently perform tasks, and self-navigate accurately to the appropri-ate location. Specifically, the drones are widely applicable in the fields of agriculture, construction sites, mining, healthcare, inventory manage-ment, and telecommunication. The overall manufacturing process (e.g., car manufacturing) is supervised by an automated control system, al-though with a minuscule human intervention.

(b) 3D printing: The advent of 3D printing has overcome the limitations of traditional production mechanisms and provides a wider scope for indus-trial designs. It is applicable in rapid prototyping and manufacturing in industries. 3D printing produces a three-dimensional solidified object un-der the supervision of computer-based programs with certain materials, which are guided in place using a digital template.

(c) Advanced robotics: In the manufacturing industries, complex tasks are as-signed to the robots. These robots communicate with each other and learn from human workers while working with them. Further, the advancement in sensors allows the robots to perform a wide range of tasks. Kuka, a Eu-ropean manufacturer, produces robots that interact with each other and automatically adjust their actions such that the unfinished products are synchronized along a line. Another industrial supplier, ABB, has designed a robot that cooperatively works with humans.

(d) New materials: The industrial revolution acts as the stepping stone for the application of materials in different industries such as steel, ceram-ics, glass, and composites. With technological progress, lighter, stronger, adaptive, and recyclable materials have come into existence. These materi-als are self-healing in nature. For example, nanomaterials such as graphene are stronger compared to steel, but thinner than the width of a human hair.

Digital trend: Internet-based technologies interconnect the various physi-cal devices, develop new products, and reorganize the traditional structure of the industries. Smart sensors sense and transmit the data in real-time. These sensors are widely used in various industrial applications, such as inventory management, delivery of goods, and remote monitoring. Bitcoin, an application of blockchain technology, is used to keep track and secure transactions. With the advent of 5G communication technologies, data are transmitted at a higher speed compared to earlier generations, support interactive multimedia, and allow voice streaming.

Biological trend: The innovations in a biological domain have facilitated in gene sequencing. Bio-engineered bacteria, which are genetically altered ones, provide desired characteristics, generally for therapeutic applications. For biological applications, synthetic circuits have been specially developed. In 2010, a bistable switch was developed at Boston University. Similarly, bio-engineered bacteria have been applied for targeted drug delivery, therapeutic protein delivery, and gene therapy – affecting the global morbidity and mortality rate[52].

3.3.2 Tipping points

Tipping points represent the impact of technologies in the near future. The major tipping points during the third industrial revolution were improved agricultural practices, transportation, and economic growth. With the evolution of Industry 4.0, business practices improved, and new business architectures/-models emerged. The principal tipping points of Industry 4.0 are as follows:

(a) Optimized lean production: Lean production system is a method to minimize the wastes, without reducing productivity. A lean manufacturing system was developed by Toyota motor corporation. The concept was initiated based on Jidoka* and Just-In-Time (JIT)† concept. The lean manufacturing system interlinks the various manufacturing units in a plant, establishes a connection among the assembly lines, logistics services, and the entire supply chain.

(b) Smart sensors: Smart sensors connect the devices, machines, and people in real-time. These sensor nodes are capable of monitoring operations, reducing costs, and alerting human of the unsafe conditions. The network of various sensors and their connected component often produce voluminous data.

(c) Enormous data available: A huge volume of data are available from the sensors in the structured, semi-structured, and unstructured forms in the manufacturing industries. Before and during the manufacturing processes, the quality and reliability of the product can be improved, depending on

---

*Jidoka is described as an intelligent automation system or the automation that involves human touch. In case of any defective products being produced, the machine stops immediately.

†The production process is continuously upgraded, which identifies and removes the various wastes associated with production.

the availability of real-time data, which controls and coordinates the process flow in a plant. Therefore, proper synchronization of the production and the supply chain activities enables the JIT process.

(d) Increasing end-users' demand: Although new technologies are gaining popularity among the end-users, they are quite expensive. For example, 3D printing has provided end-users access to customized products through mass production. In the era of the fourth industrial revolution, smart machines develop processes quickly and efficiently, which has led to an increase in demand from end-users.

(e) Analysis of data: Analysis of the data available enables business processes to run efficiently and help the machines to develop self-healing characteristics. The big data available possess certain characteristics in their nature, such as volume, variety, veracity, and complexity. Further, the predictive analysis of the data using various tools minimizes the downtime of machines, minimizes defects in the developed products, and forecasts the life cycle of machines and the necessary maintenance of finished products.

3.4   Sustainability Assessment of industries

The term "sustainability" indicates continuity at a fixed rate. The development in manufacturing industries started in the first industrial revolution. For sustainability, Industry 4.0 incorporates the technologies developed until Industry 3.0 and the emerging technologies after that [55]. The manufacturing industries form the foundation of modern industrialization and play a massive role as the keystone of the progress in the world economy. The estimation of relevant issues, performance metrics, and the range of dimensions are done in Industry 4.0. Sustainability in manufacturing industries includes specific features such as efficient usage of energy, conservation of non-renewable resources to meet the requirements of the future generation, proper usage of the renewable resources, safety and security of workers, reduced production of wastes, and usage of environment-friendly materials. In the case of developed, emerging, or developing countries, the presence of a sturdy manufacturing base is essential. Therefore, sustainability is the primary requirement of the manufacturing industries, which has the potential to bring changes worldwide. Sustainability in the Industry 4.0 perspective is discussed under two heads – globalization and socio-economic effects.

3.4.1   Globalization effects

Among the various drivers of sustainability in the manufacturing industries, globalization plays an important role. In order to evaluate the requirement of globalization, it is important to identify the necessities. Based on the principal elements, the necessary issues are updated [56]. Fig. 3.4 shows the major globalization issues to be addressed. The major challenges faced by globalization issues are discussed as follows:

(a) Energy is generated at the power stations, transmitted through the transmission lines to the substations,[‡] transformed to lower voltage, and distributed and consumed by the end-users.[§] The fundamental obstacle in the sustainable development of the manufacturing industries is the continuous availability of energy at reasonable prices. Fossil fuels are generally used to produce energy. The conservation of fossil fuels is essential for ensuring their availability and their future use. This has a remarkable impact on the economic sustainability of organizations. Therefore, renewable sources of energy, such as solar energy, biomass, wind energy, and geothermal energy, are used to meet the increasing demand and fluctuations in energy prices. To promote sustainable development, it is important to maintain energy efficiency and monitor the fluctuations in energy prices.

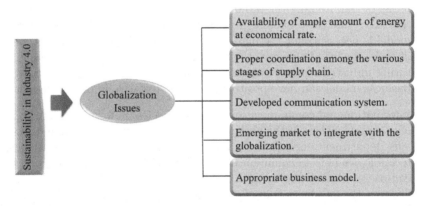

FIGURE 3.4: Globalization issues of sustainability in Industry 4.0.

(b) Supply chain management (SCM) is one of the most important aspects of the sustainability of various industries. The traditional supply chain is a loop, which consists of various entities, information, resources, and people engaged in delivering the product/service from the supplier to the consumer. Timely delivery of the product and maintaining reliability by the suppliers is important. With the advancement in technologies, smart

---

[‡]Electrical power is transmitted through either overhead transmission lines or underground cables in the form of alternating current (AC) or direct current (DC).

[§]End-users may be a domestic user or industrial user.

sensors and other technologies assist in tracking the product, smart inventory management, and maintaining the time window for product delivery. To retain the sustainability in the supply chain, environmental factors should also be considered.

(c) The emerging market has the characteristics to congregate the newly developed and innovative products. The countries with emerging markets vary from big to small. Unlike traditional markets, emerging markets follow a decentralized trend of products. The reformation of the traditional market into an emerging market leads to an increase in the Gross Domestic Product (GDP) of the entire nation. The outcome of the emerging market leads to an increase in overall profits. On the other hand, gradually, the integration between the manufacturing industries and globalization takes place. This improves the social stability and standards of living of the societal middle-class. Proper government regulations are also required to preserve the public and private sectors.

(d) Business model symbolizes the organizational and financial structure of the entire business. It depicts the plan of any organization to generate profit by maintaining the product quality and reputation of the organization. To retain the economic growth of any organization, a proper business model is necessary for the marketing, selling, and appropriate conversion of technologies. Therefore, the economic sustainability of the manufacturing industries largely depends on their business model. A sustainable business model can have positive as well as negative impact economies.

(e) Information and communication technologies (ICT) form the central part of an organization. In the absence of proper ICT, the link among the various departments or offices in industries remains absent. Internet-based technologies help in communication among the suppliers, producers, and end-users. The advancement in communication technologies indirectly influences the production in the manufacturing industries. The advanced ICT promote a decentralized market (open market). Thus, ICT form an important factor in the globalization effects.

3.4.2   Socio-economic effects

Sustainability in industries is dependent upon social, economic, and environmental effects. In the Industry 4.0 context, sustainability in the manufacturing industries is based upon socio-economic factors such as growth in per-capita income, cultural diversity, productivity, preservation of natural habitat, and proper utilization of natural resources, as shown in Fig. 3.5. Population growth also affects the socio-economic aspects of the organization. As per the economists, there is a dual perspective held by the population – pessimistic and optimistic. The pessimistic opinion provides the opinion that population rise obstructs economic growth and depletes economic investments. However,

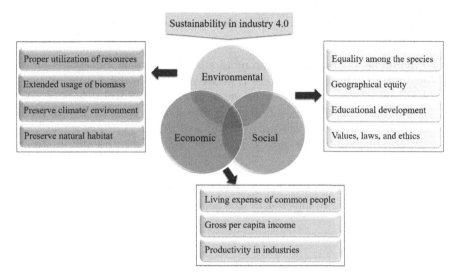

FIGURE 3.5: Socio-economic issues of sustainability in Industry 4.0.

as per the optimistic opinion, population growth will result in knowledge dissemination and improvement in education and businesses. The various socio-economic effects are as follows:

(a) Economic factors: The economic sustainability of industries indicates the cost of living of common people, gross per-capita income, and productivity in the industries. The economic growth indicates the overall qualitative growth, rather than quantitative growth. Improvement in the transportation system is also a part of the sustainable living of common masses. The economy may undergo crisis, recession, and depression. Generally, the economic crisis lasts for a few months. If an economic crisis lasts for a longer period, it results in an economic recession. The exponential fall in the economic condition leads to recession. The after-effects of recession are characterized by a fall in the GDP, actual income, employment, industrial production, and wholesale retails. Economic depression is a serious condition of economic recession, which increases unemployment, bankruptcies, and a reduction in trade and commerce.

(b) Social factors: Social sustainability signifies social equality, which affects economic growth. Social development infers the upgradation of both individual and overall well-being of the society. These developments result in the enhancement of the social capital.¶ Further, social development leads to an improvement in the educational status of the region. Human rights

---

¶To fulfill their expectations, people use certain resources and trustworthiness among them, which are termed as social factors.

and government laws alter the economy, society, as well as the environment.

(c) Environmental factors: To promote sustainability in industries, proper utilization of natural resources is necessary. A balanced ecosystem maintains population, preserves the environment, and natural habitat. Any disturbances in the ecosystem affect sustainable growth. The available natural resources are categorized into two types – (1) renewable,‖ and (2) non-renewable** resources. Thus, the usage of eco-friendly materials should be encouraged.

Therefore, the overall well-being of society and sustainability in industries is directly or indirectly affected by the social, economic, and environmental factors.

3.5 Smart Business Perspective

The business model of any organization is designed based on the content, structure, and governance of the transactions of business [57]. Fig. 3.6 shows the structure of a business model. Smart and connected products form the foundation of a smart business model. The transformation of traditional models to the smart business model has been observed to have been driven due to service-centric characteristics. These automated business models reorganize the structure of manufacturing industries.

Smart and connected products are categorized into four sections, such as to monitor, control, optimize, and connect the products developed [58], as illustrated in Fig. 3.7. The monitor, control, and optimization of the product result in the automation of the smart product. The connectivity between the products may be one-to-one,* one-to-many,† and many-to-many.‡ Connectivity permits information exchanges among the products and the external environment. Further, connectivity allows the characteristics of the product to exist

---

‖Renewable resources are the resources, which are naturally available such as solar energy, wind energy, water energy, geothermal energy, and tidal energy.

**Non-renewable resources are resources that are not replenished over time such as coal, petroleum, and natural gas.

*One-to-one connectivity indicates the connectivity among any individual product with other products.

†One-to-many connectivity indicates the connectivity among the central system to multiple products.

‡Many-to-many connectivity indicates the connectivity among the multiple products connected with external data sources.

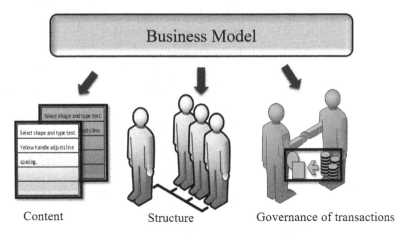

Content   Structure   Governance of transactions

FIGURE 3.6: Structure of a business model.

outside the device. The overall efficiency and productivity in manufacturing industries are improved using smart products.

(a) *Monitor*: The smart products such as smart sensors, microprocessors, control, software, processors, and smart devices are capable of observing the product quality, product operation (duty cycle of the product), and the other external conditions that affect the product in real-time. The products receive alerts from users to change their performance.

(b) *Control*: Smart products allow us to control and customize the product based on the user's experience. These products are remotely controlled through the software embedded into the product.

(c) *Optimize*: The optimization of a smart and connected product helps to improve the quality of the product, remotely operate the product, and repair it. The product is monitored with the help of certain algorithms.

(d) *Automate*: Monitoring, control, and optimization help in the autonomous operation of products. Smart products can self-diagnose faulty conditions. Further, these products operate while synchronizing with other systems.

Smart and connected products prove beneficial for the industries against various specific aspects [59] as described in Fig. 3.7, are as follows:

(a) The smart products can be monitored both remotely and in real-time. Smart sensors and other equipment have resulted in cheaper products, efficient maintenance, and repair.

(b) As the manufacturers are capable of monitoring the product continuously, the product is upgraded. This improves customers' satisfaction.

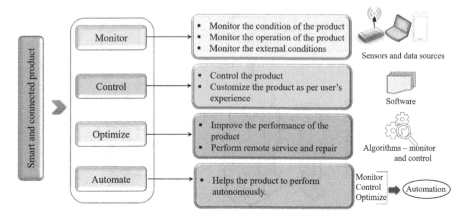

FIGURE 3.7: Smart and connected products.

(c) New business lines are created depending upon the analysis of the data received from the smart products. These data may also be helpful for other organizations.

(d) Smart and connected products can be easily reused or recycled, which has a positive impact on the environment. Improved tracking of individual products encourages reuse of the products.

(e) With the help of smart and connected products, a smart supply chain is developed. Therefore, as the product delivery and response time reduces, the just-in-time (JIT) manufacturing process is enabled.

3.5.1   Characteristics of smart business model

In the present era, the prosperity of any organization depends on the unique, summarized business statement, value proposition,* various sources of revenue streams and technologies. The value proposition is a sort of assurance and understandable statement expressed by the company to a customer, as described in Fig. 3.8. The various sources of value proposition in any organization is categorized as:

(a) Novelty-centric: Novelty in any business model denotes new transaction structures in business, new services, and new technological innovations. Digitization and integration of the technologies in Industry 4.0 result in the novelty of businesses.

---

*The convincing business statement which outlines the reasons regarding the purchase of any product or service. Further, this statement illustrates the reason for the product or service being added such that it adds more value than another product.

(b) Efficiency-centric: Smart and connected businesses enhance the efficiency of an industry. The improvement in transaction efficiency leads to faster, simpler, less error-prone, and increased dependability of transactions.

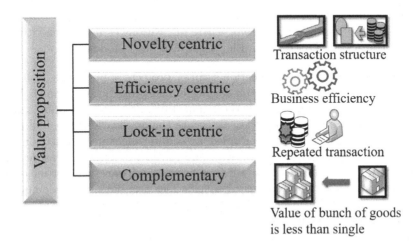

FIGURE 3.8: Value proposition of smart business model.

(c) Lock-in centric: Lock-in encourages customers for repeated transactions. At the same time, lock-in prevents the migration of customers. Further, trust is also escalated among the buyers and sellers.

(d) Complementary: Complementaries in the business are present whenever a bunch of goods/services are of higher value compared to the individual value of that product. Data also provide a detailed description of the online and off-line assets. Smart businesses also complement technologies.

3.6 Cybersecurity

The various protection measures that are undertaken to secure internet- connected systems – software, hardware, and data – from malicious attacks are commonly known as *cybersecurity* measures. To avoid unauthorized access of data, organizations rely on *cybersecurity* and *physical security*. These attacks usually access, alter, and destroy sensitive data from the data centers.

Additionally, these malicious attacks act as interruptions to the normal execution of the business process. In the era of the fourth industrial revolution, millions of devices are connected through Internet-based technologies. Therefore, these devices and the data transmitted using these devices are more prone to attacks. The cyberattacks may be at the individual or organization level. At the individual level, any attack may lead to identity theft, which results in the loss of crucial data of that person. While at the organization level, the loss or alteration in the critical data of various shop floors may interrupt the operation of the overall system. There are various essential constituents of cybersecurity, as illustrated in Fig. 3.9, which are as follows:

(a) Application Security: Application security protects applications from vulnerable threats, using software, hardware, and procedural methods. After the software development, as the applications are available online, they become more prone to attacks. To secure the applications, specific measures are taken, which are termed as counter-measures. Based on the type of security required, software and hardware counter-measures are applied. An application firewall is a form of software counter-measure, which provides limited access to specific files and data of the computer. Further, the application firewall does not allow specific applications to access the operating system of the computer. The commonly available hardware counter-measure is a router, which blocks the IP address of the computer from being visible and exposed on the internet.

(b) Information security: Information security represents specific strategies that control the processes, policies, and tools required to prevent digital and non-digital threats. The various information or data are protected irrespective of the format of information. The information may be in transit or at the resting stage in storage. Generally, information security maintains the confidentiality,[†] integrity,[‡] and availability[§] of the information or data. This is also known as the CIA triad. The sensitive information may suffer from several threats such as malware and phishing attacks, and identity theft. To prevent attackers and reduce vulnerabilities at different points, a layered defense in depth strategy is undertaken.

(c) Network security: Network security represents the set of policies that monitor and prevent unauthorized access. Additionally, network security safeguards the network and data from the network traffic. The primary objective of network security is to prevent various threats from entering the

---

[†]To avoid the tampering of sensitive information, they are revealed only with certain people, which is known as confidentiality.

[‡]Integrity prohibits illegal access of data, which maintains consistency, accuracy, and trustworthiness of data.

[§]Availability of data indicates the reachability of data to the authorized persons.

network. It incorporates multiple layers of defense, which are mostly applicable at the edge and in the network. There exist two types of attacks – active and passive. In active attacks, the network intruder/attacker gives commands to distort the regular operation of the network. For example, eavesdropping, masquerade attack, session replay attack, and denial-of-service attack are different types of active attacks on the network. While in the passive attacks, the intruder monitors, reads, and analyzes the data traveling through the network. For example – wiretapping, idle scan, and port scanner are different types of passive attacks on the network.

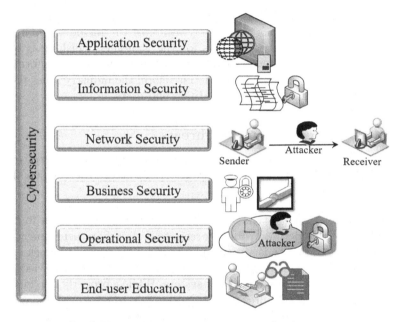

FIGURE 3.9: Various constituents of cybersecurity.

(d) Business security: Business Continuity Planning (BCP) is a step-by-step cyclic approach used to recognize and prevent threats regarding any unplanned incident. Since it is a structured approach, the after-effects of a security incident can be minimized. BCP expresses the necessary functions of the business and identifies the sustainable processes and their details. All forms of business disruption are taken into consideration. In any organization, a proper BCP safeguards the reputation and reduces the risks occurring due to cyberattacks or any natural disaster. BCP should comprise of certain constituents such as initial data, revision management process, policy information, purpose and scope of the plan, step-by-step procedure, flow diagrams, and schedules to upgrade the plan.

(e) Operational security: Operational Security (OpSec) is a logical risk management process, which helps the managers to identify and protect critical information from the attackers. The five steps of OpSec are as follows – (1) identifying the sensitive or critical data, (2) identifying the potential threats, (3) analyzing the loopholes, (4) risk evaluation, and (5) application of appropriate measures to counteract the threats. An OpSec program is optimally implemented as an exact change in the management process, regulates the access of information, provides minimum access to the employees, and automates the tasks with minimum or no human intervention.

(f) End-user education: End-users provide the biggest threat to the industry. As the end-users are not completely aware of the intruders, sometimes they unknowingly open the doors for the attackers. Phishing is one of the commonly known cybersecurity risks, where the critical information of an organization is compromised. Therefore, the industry must provide knowledge regarding cyberattacks to its end-users.

3.6.1   Various cybersecurity threats

There are various upcoming technologies, security trends, and threat intelligence, which must be safeguarded from different forms of cyber threats. These are as follows:

(a) Ransomware: This is a type of malicious software. The attacker locks the computer of the victim, or limits access to certain files, using encryption techniques, and demands a ransom amount to decrypt and unlock access. In case of a simple ransomware attack, a tech-savvy or knowledgeable person may be able to unlock their system. While a properly implemented ransomware attack requires the decryption key to recover the files, most of the commonly used ransomware attacks use *Trojan*, where the user receives a legal file as an attachment and is skillfully forced to download or open it.

(b) Malware: Malware is a software that secretly affects the system without the user's consent. Mostly malware affects a standalone computer or a networked computer. The primary goal of malware is to steal data, important credentials, interrupt the normal operation, and demand ransom amount from the victim. Malware is observed in the form of a Trojan horse, virus, or worm. The type of malware is differentiated based on the design of the malware, how the malware is delivered to the target, and the way it avoids being detected.

(c) Social Engineering: Social Engineering is the term used in the context of information security. Attackers strategically trick users into disclosing sensitive information, which is generally protected. In other words, social engineering psychologically manipulates the user. The life cycle of social engineering involves the following steps – identification and investigation of information regarding the victim, engaging the victim and defrauding them, extracting information from the victim, and removing all the malware traces. The common social engineering attack techniques include baiting, scareware, pretexting, and watering hole.

(d) Phishing: Phishing is one of the most common forms of cyberattacks, where false information is provided to the end-users, through emails that resemble being sent from reputed sources. The primary aim of the intruder is to obtain user data such as login credentials and credit card details. The victim is deceived into clicking on a link, which leads to malware installation in the user's computer. Further, this leads to the disclosure of sensitive information. Phishing is of two types – email phishing scams and spear phishing. In the case of email phishing, the attackers send thousands of deceitful emails that resemble actual emails from credible sources while spear phishing attacks are targeted to a specific person or organization to obtain sensitive information. To protect sensitive information, users may check with false passwords/credentials unless they are sure about the website.

3.6.2 Requirements of cybersecurity

Cybersecurity is one of the most critical roles in various industrial sectors. The traditional cybersecurity architectures have specific security characteristics such as confidentiality, integrity, non-repudiation, and access control [60]. These security mechanisms obstruct computer/network intrusions and attacks. The modern cyberattacks are vast, continually evolving over a given period, and are complex compared to the traditional ones. Algorithms collect and analyze the data generated by various events in a cyberattack environment. Based on the data collected, these algorithms intent to accurately detect suspicious activities. These detection systems form the core of the attack detection system. However, in terms of accuracy, the objective of the algorithms is hard to achieve. The detection algorithm, which gives incorrect results, negatively affects the performance of the system.

The cloud-based design and advanced manufacturing techniques establish a relationship between the IIoT and the Internet of Services* to develop collaborative cyber-physical products. Software-defined cloud manufacturing architecture (SDCMA) is a hybrid decentralized system. There are three parts of the SDCMA system – software plane, hardware plane, and ensemble intelligence framework. The software plane consists of the virtual layer and the

---

*Internet of Services provides the users a decentralized way to exchange online services and goods. It also enables organizations to develop applications for supporting a large number of users.

control layer. To perform any specific task, the communications are managed by the control elements (CE) present in the control layer. Further, the CE also provides data to the ensemble intelligence layer. The design and manufacturing works are performed in the distributed hardware layer present in the hardware plane.

Computationally Intelligent Systems (CIS) deal with huge volumes of numerical data, along with uniform advancing and error rates, which replicate human performance [62]. CIS requires an algorithm, which combines and filters data generated in the various events of the cyber domain. During the process of decision-making, CIS systems are designed to handle high dimensional and noisy data. CIS systems provide preferences to construct new categorization algorithms for the detection system. It is also referred to as the subset of AI.

3.7   Impacts of Industry 4.0

With the advent of Industry 4.0, substantial changes have been observed in the technologies and industrial infrastructure [54]. The digital and physical world gets connected through the Internet. The real-time data collected from the entire process help in improving the efficiency of the process, reduction in errors, and quality of the product. Industry 4.0 has brought a wave of transformation in the manufacturing industries by varying the workforce and systems. These changes significantly affect the *economy*, *business*, and *global perspective*, which are as follows:

3.7.1   Economy perspective

The physical, digital, and biological trends of Industry 4.0 have led to the development of new technologies and improvement of the overall system. Further, this technological development results in the changes in the lifestyle of common people and controls the digital infrastructure of the system. With the inclusion of digital trends, citizens get a public platform to voice their opinion, coordinate their efforts collectively, and receive updates of the government activities and norms. The competitiveness among public organizations also increases. Therefore, the outcome of new technologies results in the redistribution and decentralization of power, as shown in Fig. 3.1.

3.7.2   Business perspective

The sales, marketing, and distribution of the digitized business will improve compared to the traditional business methods. Industry 4.0 results in the

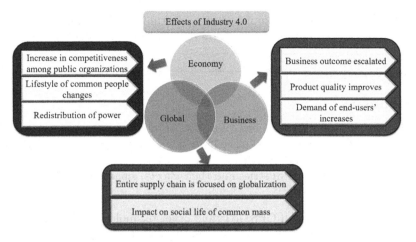

FIGURE 3.10: The outcomes of Industry 4.0.

introduction of new technologies and accelerates innovations, and improvement in business. The fourth industrial revolution will result in improved productivity, efficient usage of resources, and efficient production. As the percentage of defective products produced reduces, the quality of the product improves, thereby the business outcome has escalated. With the increase in the demand of end-users, the organizations are bound to adopt the advanced technologies to deliver the products to the market. From the business perspective, the principal effects of Industry 4.0 are – increase in the expectation of customers, improvement of the product, and collaborative innovations. Fig. 3.1 illustrates the various business effects of Industry 4.0.

3.7.3 Global perspective

The industrialization has already influenced the businesses to flourish globally. With the advancement in transportation and availability of the 5th generation $(5G)$ communication, overseas trade has increased. The entire supply chain is centrally focused on globalization. Goods are sent across the globe by airlines, ships, and trains, daily. The analysis of real-time data and digital labeling and digitization of products are the key attributes of Industry 4.0. Further, the tracking of the goods, logistics, managing the facilities, services, and maintenance have global impacts. The technological improvements impact the social life of common people. For example, smartphones tend to hold back people from meaningful interaction, and rather the upcoming generation is more engaged with the virtual applications and interactions on the phone.

Almost all developing countries adopt new business models and value-producing opportunities to implement digital technologies. Digitalization is the preliminary step to enforce the other advanced technologies of Industry

4.0. The creation of new employment opportunities, new business models, reusing, and recycling the resources are the outcomes of Industry 4.0.

Summary

This chapter comprises the basic concepts of Industry 4.0—historical perspective, significant changes in the industry, design requirements, and the drivers of Industry 4.0. Further, this chapter also discusses the sustainability assessment of industries and their effects, smart business perspective, and cybersecurity.

(a) The combination of CPS, IoT, and other technologies led to the evolution of the fourth industrial revolution. Industry 4.0 is another name for the fourth industrial revolution.

(b) Industrial revolution has many adverse effects, such as the depletion of natural resources, an increase in the level of pollution, child labor, and the exploitation of laborers.

(c) Information clarity, interoperability among the devices, machines capable of taking their own decisions, and technically advanced systems are specific design requirements of Industry 4.0.

(d) Safety and security of workers are significant aspects of Industry 4.0.

(e) Technological advancements and the upcoming impact of these technologies are the drivers of Industry 4.0.

(f) Sustainability in the industries has two facets – globalization and socio-economic effects.

(g) Smart and connected products form the base of smart business. Smart business improves the efficiency of a business, enhances new transaction structures in businesses, and encourages customers to make repeated transactions.

(h) To avoid unauthorized access to sensitive information, various security measures are undertaken by organizations, known as cybersecurity. Ransomware, malware, social engineering, and phishing are some of the common forms of cyberattacks.

(i) Industry 4.0 has brought substantial changes in the industries, which can be classified from an economical, business, and global perspective.

Exercises

1. What are the impacts of the industrial revolution?

2. Describe the design requirements of Industry 4.0.

3. Define the drivers of Industry 4.0. Discuss the different drivers of Industry 4.0.

4. Which technical innovations form the physical technical driver of Industry 4.0?

5. What are the principal tipping points of Industry 4.0?

6. What are the features of sustainability in manufacturing industries?

7. What are the primary factors behind globalization in manufacturing industries?

8. What are the socio-economic factors of sustainability in Industry 4.0?

9. What form the basic structure of a business model? How does a smart business model is formed?

10. What is cybersecurity defined? What are the primary constituents of cybersecurity?

11. Discuss the significant impacts of Industry 4.0.

4

Industrial Internet of Things: Basics

Learning Outcomes

■ This chapter covers the basics of the Industrial Internet of Things (IIoT), differences between traditional automation and IIoT, and the Industrial Internet.

■ New readers will be able to gain knowledge of the various aspects of IIoT, design considerations of the Industrial Internet, and its impact.

■ This chapter also provides insights regarding the Industrial Internet Consortium (IIC) and the testbeds developed by them.

■ Smart sensor nodes and their applications are also discussed in this chapter.

■ This chapter explains the Industrial Processes and their features, future architecture of an industrial plant, and viewpoints.

4.1 Introduction

Industrial Automation and Control Systems (IACS) refers to a set of systematic economic activities that results in the specific variation of the operational parameters for machines and systems, with minimum or no human intervention, giving specific signals as input. Further, IACSs are extensively used in different industries such as manufacturing, transportation, supply chain and logistics, and packaging [63]. On the other hand, security is one of the major issues in IACS systems. The insecurities or vulnerabilities of IACS systems may result in severe incidents such as loss of life, safety, lost production, and damage to pieces of equipment. For example, IT systems are prone to attacks by different viruses. Further, remote access to industrial processes may expose the systems to further attacks. The integration of automation and control across the industries help to connect site, facilities, and people. Internet of Things (IoT) depicts a scenario of connected objects or things, which communicate with each other and are identified by a unique identifier. As discussed in Chapter 2, IoT forms the base for the Industry 4.0, Industrial Internet systems, and Industrial Internet of Things (IIoT). With the development of IIoT, major changes are expected in the architecture of IACS. The key objectives of IIoT are – improved working conditions, better product quality, minimization of energy consumption, enhancement of machine lifetime, and upgradation of operational efficiencies. In a smart factory environment, the sensed data is transformed into meaningful information, and the operations are further improved using advanced analytics. As illustrated in Fig. 4.1, the data generated from the smart sensor nodes are processed, and operations are improved based on the feedback. Table 4.1 outlines the differences between IIoT and traditional automation.

IIoT architecture is described as a three-layered infrastructure consisting of – device, gateway, and platform/middleware layer. The device layer comprises heterogeneous type of smart sensor nodes which are deployed at various machines and devices. The sensor nodes sense and transmit data to the middleware layer through the gateway devices present at the gateway layer. After that, the gateway devices connect to the higher layers through infrastructures such as Wi-Fi and LAN, for further processing. The edge layer and cloud together form the middleware layer and satisfy the requirements of analysis, storage, and processes data [64]. For example, GE Predix, Siemens Mind-Sphere, and Honeywell are some of the industrial cloud platform providers. On the other hand, C3IoT, Uptake, and Meshify are some of the software development firms. Further, DTS, Honeywell, Omega, Bosch, and Adafruit are certain sensor manufacturers. Some actuator technology manufacturers are Eckart, SKF, Fuyu, Knr, and Sirius.

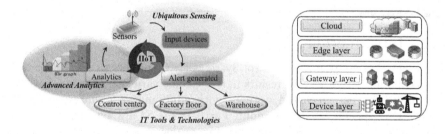

FIGURE 4.1: Industrial Internet of Things.

Definition of IIoT

The Industrial Internet of Things (Industrial IoT) is made up of a multitude of devices connected by communication software. The resulting systems, and even the individual devices that comprise it, can monitor, collect, exchange, analyze, and instantly act on information to change their behavior or their environment intelligently – all without human intervention.[65]

4.1.1   IIoT and Industry 4.0

The terms Industry 4.0 and IIoT maybe sometimes interchangeably used in the academia and industries. However, the concept of IIoT and Industry 4.0 is not the same. As discussed in Chapter 2, the primary focus of Industry 4.0 is manufacturing industries. The various automation technologies of Industry 4.0 use self-optimization, self-configuration, self-diagnosis, and intelligently support workers. On the other hand, IIoT focuses on the industrial sector as a whole. In addition to the development of a smart factory, IIoT transforms the entire business processes from maintenance, production, and order, to delivery of the product. For example, warehouse sensor nodes provide details of stock, which prevents understocking and overstocking of essential items. *Industrial Internet Consortium* (IIC) coined the term IIoT. On the other hand, the term *Industry 4.0* originated from a German Federal Ministry of Education and Research Initiative. The various automation technologies associated with IIoT mostly depend on intelligent CPS. In late 2012, General Electric coined the term Industrial Internet. Both paradigms of Industry 4.0 and IIoT are associated with the generation of a colossal amount of data, processing, and transmission of these data in real-time. In IIoT, the middleware layer should be scalable, reliable, able to handle fast computations and be responsible for securing the whole system. However, any failure in the system may lead to emergencies, which may be detrimental to infrastructure or even prove fatal to human lives.

4.1.2   IIC

IIC is an open membership organization, founded in March 2014 [101] to expedite the growth of trustworthy IIoT systems. IIC was founded by AT & T, Cisco, General Electric, IBM, and Intel. However, the parent founder company of IIC is Object Management Group. IIC is a non-profit organization created to promote open standard technologies and interoperable technologies. IIC collaboratively works together with various organizations to shape future IIoT and make innovative ideas feasible. The main objectives of the IIC are

Table 4.1: Comparison between IIoT and traditional automation.

Features	IIoT	Traditional automation
Real-time communication	M2M communication is possible with minimum latency (dynamic). Better access and usage of the sensor nodes data.	Initially started with relay logic, with ladders and rungs. The development of Programmable Logic Controller (PLC) was a quantum leap in this direction, which replaced the relays and rungs.
Data model	Point-to-point data model	Broadcast-subscriber model
Data usage	The received data are translated into actionable information to improve operations, receive feedback, and execute accordingly.	To improve the operations, data are not directly translated.
Visualization	In order to optimize the entire production process, it can be visualized in a better transparent way.	Visualization of the entire production chain is semi-transparent.

as follows:

(i) To drive innovations toward the design of new testbeds and use cases for real-world applications.

(ii) To promote interoperability, new reference architectures, and development of frameworks.

(iii) To instigate a global standard process for Internet and industrial systems.

(iv) To share new ideas, practices, and provide insights through an open forum.

(v) To frame up new and innovative ideas for the security of society.

IIC testbeds comprise details of industrial assets, collect status data, improve the quality of collected data, and create a network of smart factories [101]. The primary goal of IIC is to exhibit the Industrial Internet solutions to real-world implementations. IIC released ten public testbeds in February 2016, which are listed as follows:

Track and Trace: This testbed introduces the Industrial Internet to the different units of the factory. The primary goal is to regulate the different instruments in the manufacturing and maintenance environment.

Communication & Control Testbed for Microgrid Applications: This testbed proposes to remodel the traditional power grid into multiple microgrids. Microgrids assist in the integration of renewable energy sources and prevent a power outage.

Asset Efficiency Testbed: This testbed efficiently gathers accurate real-time information and use predictive analytics to make appropriate decisions. The testbed is cast in two phases – the first phase created the moving solution, and the second phase addresses the fixed assets.

Edge Intelligence Testbed: This testbed escalates the development of the edge architecture and algorithm by providing access to the developers to a wide range of hardware and software systems at a low cost. The applications of IoT in industries require real-time analysis of data at the edge of the network, which is provided by this testbed.

Factory Operations Visibility & Intelligence Testbed (FOVI): This testbed provides a simulated factory environment. The FOVI helps to visualize how a process is optimized, how to process the real-time data, and use analytics.

High-Speed Network Infrastructure Testbed: This testbed introduces high-speed fiber-optic lines to industrial systems. This is designed to enable the Industrial Internet in the factory units. The high-speed network transfers data at a speed of 100 gigabytes per second (approximately). A seamless M2M communication and data transfer across the network is made possible because of this high data transfer speed.

Industrial Digital Thread Testbed (IDT): This testbed improves the efficiency, flexibility, and speed of industrial processes, through digitization and automation of manufacturing processes. The smart sensor nodes in industries help to optimize the operation and flow process in the supply chain.

INFINITE Testbed: The International Future Industrial Internet Testbed (INFINITE) assists in the development of the infrastructure of Industrial Internet products and services. This testbed utilizes big data to connect multiple virtual networks via a physical network.

Condition Monitoring and Predictive Maintenance Testbed (CM/PM): This testbed provides the indications of early failure or degradation in the performance of the machines through continuous monitoring of the system using sensor nodes. Further, modern analytics help to recommend corrective measures to detect and rectify the problem.

Smart Airline Baggage Management Testbed: This testbed provides a better aviation system that will reduce the cases of delayed, damaged, and lost items of luggage, and increase the ability to track luggage and check variation in the weight of the luggage. Additionally, this testbed aims to improve the customer's satisfaction and provide new value-added services to frequent flyers.

4.2   Industrial Internet Systems

As per [67], Industrial Internet Systems are *the connection of industrial machine sensor nodes and actuators to local processing and the Internet; the onward connection to other important industrial networks that can independently generate value.* The convergence of the global industrial system with advanced computing methods, analytics, smart sensing, and new connectivity through the Internet has led to the emergence of the Industrial Internet. Fig. 4.2 illustrates the essential elements of Industrial Internet – intelligent and automated machines, interconnected people at work for providing support to the intelligent operations, service and safety to people, and advanced predictive and prescriptive analytics. The Industrial Internet systems interconnect the industrial machines and incorporate the other industrial networks. This interconnection improves the operational efficiency of the machines. Further, the efficiency of the entire industrial network is upgraded. Therefore, the overall performance, uptime of the machines, and productivity in general rise.

Industries typically have heterogeneous devices that are interconnected and generate a massive volume of data. The sensor nodes deployed at the machines sense and transmit data to the edge of the network and cloud for processing, analysis, and storage of the generated data. The storage, processing, and computation of the massive amount of data managed using big data technologies. Various advanced technologies, such as machine learning and deep learning, are used for the analysis of data. The outcome of analytics provides meaningful operational insight – predictive maintenance, prediction of faults, optimal fleet operation, improvement in product quality, and reduction of overall costs. For example, in the healthcare sector, a tomography scanner is used to visualize the image of the internal body parts. These scanners detect and treat various healthcare afflictions associated with organs and parts, such as the heart, chest, brain, and abdomen. On the other hand, in oil refineries and petrochemical plants, operators monitor and model the various rotating machines such as reciprocating and centrifugal compressors, for predictive maintenance and safety. As a result of the emergence of Industrial Internet Systems, there has been a substantial improvement in the income and living standard of workers and other employees of these industries.

4.2.1   Design of industrial internet systems

Smart sensor nodes and advanced technologies form the basic building block of the Industrial Internet System that interconnects the machines. Sensor nodes are deployed at various locations in the industries to connect various disjoint industrial infrastructure and equipment/machines. The huge amount of data collected from these sensor nodes are processed, and the feedback received by them help to improve the performance of the machines. The continuous effort

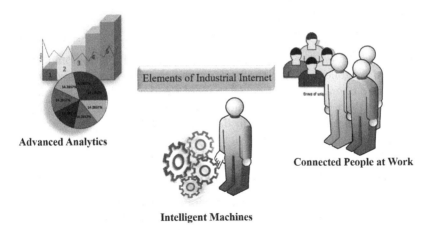

FIGURE 4.2: Essential elements of Industrial Internet.

of individuals combined with the investment done to deploy the sensor nodes acts as the key enabler of the Industrial Internet systems. Additionally, the robust security system, data scientists, and engineers play the role of catalyst in the development of Industrial Internet systems [68]. The early adoption of industrial Internet-based technologies also leads to the all-round development of the industrial sector of developing nations. Further, this industrial development will bridge the gap between developing and developed countries. The key factors in the design of Industrial Internet systems are discussed as follows:

i *Intelligent Machines*: The smart sensor nodes, control systems, and advanced applications form the basic part of the connected infrastructure of the Industrial Internet. The data obtained from the sensor nodes help to gear up the intelligence of the system. Because of the use of improved microprocessor chips, the computing power of these machines has improved. Further, *intelligent systems* can be developed with the help of intelligent machines. The intelligent systems may be represented in different forms, such as network optimization, maintenance optimization, learning, and system recovery.

ii *Connected People*: Connected people denotes the interconnection between people working at various industrial sectors, or moving at different locations. These connected people help in the design, operation, and maintenance of the system. For example, in the healthcare sector, a doctor can virtually interact with his/her patients, and the nurse or paramedic can remotely monitor the patient's conditions. On the other hand, in industries, managers can remotely monitor a unit's condition. This results in the increased safety and security of the worker and enhances the quality of service (QoS) of these industries.

iii *Advanced Analytics*: The predictive algorithms, automation, and analysis of the data collected from the sensor nodes result in the improvement of a machine's operation. Further, analysis of the data enhances the efficiency of the system. The data obtained from these intelligent devices are collected and processed to help in data-driven learning. This results in the development of intelligent decision-making in Industrial Internet systems.

4.2.2   Impact of industrial internet

In order to scale the industrial systems globally, the role and the opportunities created by the Industrial Internet are appreciable. The industrial internet has a widespread impact, which is challenging to quantify in uniform terms. Thus, the impact of Industrial Internet can be described through three different perspectives: *economic, energy consumption*, and *physical assets*. These three perspectives together describe the usefulness and scope of the Industrial Internet on a vast scale [68]. Fig. 4.3 illustrates the three different perspectives on the impact of the Industrial Internet and briefly describes each of them. These perspectives are elaborated as follows:

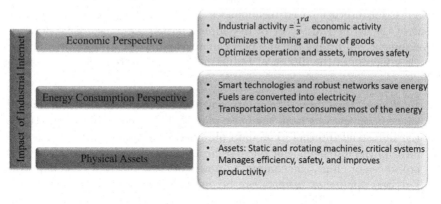

FIGURE 4.3: Impact of Industrial Internet.

4.2.2.1   Economic perspective

The traditional economic measures vary from one country to another. However, industrial activities approximately form a third part of the economic activities of a nation. The Industrial Internet comprises huge number of sectors such as manufacturing, natural resource extraction, and construction.

Further, economic activity also includes a large portion of the transportation sector, such as aviation, rail, and marine transport. The entire supply chain management is dependent upon the movement of goods, services, and information. Industrial Internet optimizes both the delivery and flow time of goods in the heavy industries. The economic effects of the Industrial Internet will also have a positive impact on the healthcare sector. Therefore, improvement in the different industrial sectors will prove to be beneficial for overall economic development.

4.2.2.2 Energy consumption perspective

The evolution of smart technologies and their integration with the existing technologies result in efficient energy savings and reduce costs. The major problems in the present-day world are scarcity of non-renewable resources, the requirement of environmental sustainability, and the absence of proper and sustainable infrastructure. The generation of electrical energy involves a wide range of activities, which include extraction of fossil fuels, refining and processing the fuels, and finally converting the fuels into electricity. Approximately more than half of the world's energy is consumed by the manufacturing and transportation sectors. The large-scale adoption and deployment of Industrial Internet in manufacturing industries can cause a reduction in the energy consumption through process integration, life-cycle optimization, efficient utilization, and maintenance of static and rotating parts of machines. In order to reduce the amount of energy consumption across different industrial sectors, there is a need to explore the performance of each of these systems and their associated subsystems. Therefore, the growth of the Industrial Internet improves the process of energy production and conversion.

4.2.2.3 Physical assets

The physical assets of an industrial system comprise of static and rotating machines, and complex systems. Each of the machine parts provides meaningful information, which is important to monitor the performance of the unit and the system as a whole. To scale any industrial system, it is necessary to monitor certain specific units. It is not very easy to know the details of the existing devices, systems, and networks in different sectors such as transportation, oil and gas, power plants, industrial facilities, and healthcare. On the other hand, the physical assets of such industries are exposed to variations in temperature, vibration, pressure, and other metrics that can be remotely monitored, manipulated, and modeled to improve safety, productivity, and efficiency [68]. Fig. 4.4 illustrates the examples of different physical assets of power plants such as hydroelectric, steam, nuclear, geothermal, and diesel-electric power plant. These physical assets comprise static and rotating parts of the power plant.

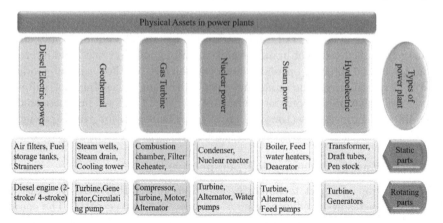

FIGURE 4.4: Physical assets of a power plant.

4.2.3   Benefits of industrial internet

The Industrial Internet proved to be beneficial for different machines, industrial networks, supply chains, and the global economy. The global industrial system encompasses different industries, supply chains, facilities, and networks. Therefore, a small improvement in the efficiency of any sector boosts the performance of the whole system. The various industry-related benefits of the Industrial Internet are discussed as follows:

1. With the widespread usage of the Industrial Internet, there has been positive growth in the overall production of goods and services, which also has a big effect on the global economy.

2. Intelligent sensor nodes and other technological advancements result in the optimization of the performance parameters of each machine in the industries.

3. As the sensing and data handling costs of the intelligent devices reduce at the product level, a wide-range integration of the systems and subsystems is occurring.

4. The tracking of products, proper coordination among the workers, information, and goods in the supply chain, and product quality have intensified across diverse geographical regions.

5. The continuous processing and analysis of real-time data help to improve the design of new products and services.

4.3   Industrial sensing

Smart sensor nodes form the base of automation of the industrial systems. These sensor nodes continuously monitor and control the industrial processes. The outcome of the data sensed by these smart sensor nodes may be either in the analog or digital form. The various processes controlling industrial applications require close monitoring of the variables such as pressure, temperature, flow, and level. Depending on the data sensed by the sensor nodes, the *actuators* control the system. Primarily, IACS used sensor nodes to reduce the scrap* rate and enhance productivity. In industries, sensing is necessary to improve safety, quality, productivity, and reduce downtime of machines. To meet the industrial standards, low cost and reliable sensing technologies and methodologies are required.

4.3.1   Traditional sensing

Under traditional sensing methods, the feedback from the industrial processes was used for automation in the industrial control systems. Based on the feedback provided, the properties of the product could be identified, and proper measures could be taken to improve the product [69] further. The various processes and machines on the factory floor are handled with the help of control systems and information technologies, which minimizes human involvement in the processes. As the chances of failure are low, the maintenance cost of these machines remained comparatively low. Robots are programmed and deployed to perform tasks in hazardous conditions, which results in the improvement of the safety of workers. The healthcare costs, paid leave, holidays, and other employee benefits related to the human operator are also reduced significantly. As a consequence, it results in larger returns for the industries and reduces production costs. Additionally, the overall productivity, quality of the product, and accuracy of the information are enhanced. However, the initial cost of automation in the production line is very high.

4.3.2   Contemporary sensing

Here, smart sensor nodes connect to the Internet, transmit the collected data in real-time, which is analyzed at the cloud and edge of the network. Further, the outcome of the analysis of the data provides information about product lifetimes, loop efficiencies, safety condition improvements, and reliability of the product. Therefore, various industrial processes continuously monitored by these sensor nodes, result in the maintenance of the optimum conditions of

---

*In industries, the scrap is produced from the combination of materials, which can be recycled or transformed.

the processes. Wireless technologies are an important part of the smart sensor nodes. These wireless technologies help in fixing sensors at the appropriate positions avoiding the complications of the deployment of the sensor node. Several types of research have been undertaken to produce cheap and easy to use sensor nodes.

Typically, microprocessors form the core of the smart sensor nodes. These smart sensor nodes minimize power consumption by automatically storing or deleting data at periodic time intervals. One of the most critical facts of these smart sensor nodes is that the size of the sensor nodes is quite small compared to the standard ones. These sensor nodes are capable of handling huge amounts of data instantaneously. Under normal conditions, the data sensed by the sensor nodes are primarily processed in the microprocessor, which through advanced computation, either store or clear the data [70].

Definition of Smart Sensor

Sensor with small memory and standardized physical connection to enable communication with the processor and data network. – IEEE 1451 standard [73]

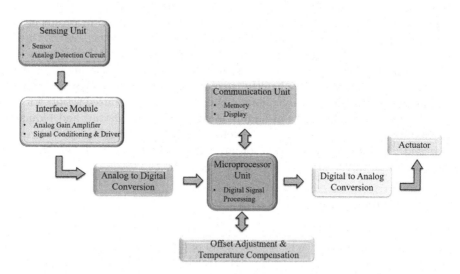

FIGURE 4.5: Block diagram of a smart sensor.

In other words, a sensor node providing a specific output when interfaced with the hardware is known as a smart sensor node. Smart sensor nodes can automatically range and calibrate data through a system, auto-correct errors, restrain fluctuations in temperature, auto-tune algorithms, and communicate through a serial bus. Fig. 4.5 shows the block diagrammatic representation of the workflow, typically followed in a smart sensor node. These smart sensor nodes transform the real-time data into a stream for transmission through a gateway. As shown in Fig. 4.5, smart sensor nodes generally comprises following components – *sensing unit, interface module, processor unit,* and *communication unit.* The microprocessor unit is capable of filtering, compensating, and processing specific signal conditioning tasks. The smart sensor nodes are also known as a system-on-chip because these devices have the electronics and transducer elements combined on a single Silicon (Si) wafer. The key application areas of smart sensor nodes in various industries are shown in Fig. 4.6. The various types of sensor nodes applicable across the industries for different applications are discussed as follows:

1. Industrial: The improvement in productivity rate, product quality and reliability, safety and security of workers depends upon the performance of sensor nodes. The most commonly used sensor nodes in manufacturing industries are temperature, pressure, level, and flow measurement, as shown in Fig. 4.6. These sensor nodes continuously monitor and control the industrial processes. Further, the utility of industrial process control is expected to follow an increasing annual growth rate over the forthcoming years.

2. Healthcare: With the advancement of technologies, intelligent sensor nodes are being used to continuously monitor the health conditions of the patients and alert them regarding any change and the onset of criticality in their conditions. The commonly used type of healthcare sensor nodes is – implantable type, biosensor nodes, and nanosensor nodes. As described in Fig. 4.6, ECG, EMG, SpO_2, and GSR are some of the common types of healthcare sensor nodes.

3. Telecommunication: In the existing GSM cellular phones, a smart card known as Wireless Identity Module (WIM) is used to provide security to e-commerce applications, using encryption and digital signatures.

4. Supply chain and Logistics: To remotely monitor and track the logistics and related activities, various sensor nodes are used, such as temperature sensors, proximity sensors, vibration sensors, and RFID tags. This use of smart sensor nodes result in reduced time to market, optimizes the supply lines, avoids theft or misplacement of items, and increases the overall efficiency of the supply chain.

5. Agriculture: Smart or intelligent sensor nodes are used to monitor and automate the agricultural processes remotely, accurately advise on the amounts of fertilizer/water/pesticide to be applied, and determine crop

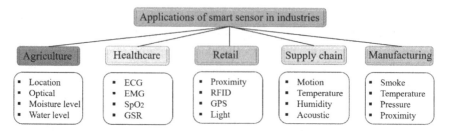

FIGURE 4.6: Applications of sensor nodes in industries.

health. Further, sensor nodes are also used to determine the best time for cropping. The various types of sensor nodes used in agricultural applications are – position, optical, moisture level, and water level.

The different industrial sensor manufacturers are Adafruit, Bosch, Honeywell, Omega, Massa, and DTS. In the industries, the GPS sensor is used for navigation purposes, such as predicting the location of the driver, analyzing traffic, and tracking objects in real-time. Further, in the agricultural industry, Agricultural Robots (Agbots) are used for introducing automation in the agricultural processes. Similarly, in the healthcare industry, systems such as a smart bed can be used to monitor the movement of the patient.

4.4   Industrial Processes

In the heavy industries, *industrial processes* form the basic core component of operations. These industrial processes are the methods which involve various physical, chemical, mechanical, or electrical stages to manufacture a product, which usually occurs on a large-scale. With the inclusion of advanced technologies in the industrial processes, IIoT forecasts the improvement of quality, safety, and productivity. As per a survey by PWC [72], the rate of process advancements is predicted to improve more than 70% by the year 2020 in various industrial sectors such as electronics, manufacturing, construction, chemical, defense, and transportation. With the completion of the transformation process, a typical industrial enterprise will be transformed into digital enterprises. Physical assets form the core of the industries, whereas the digital interface and real-time data-based services surround them. The digital transformation in the processes primarily aims to transform individual companies and the market dynamics across a wide range of industries.

4.4.1 Features of IIoT for industrial processes

The digitization of an end-to-end industrial process includes the integration of new technologies with the existing processes, industrial software and automation, and extend the communication network. The established industrial sector is expected to be reshaped fundamentally with the inclusion of advanced technologies. The key features of the industrial processes are as follows:

i Distributed decision generation for software and hardware: Smart and intelligent devices can generate their own decisions autonomously. However, in exceptional cases, interruptions, or any disagreement, the devices pass the decision to their higher authority.

ii Clarity of information to workers: The transparency of information to the operators is necessary for the industrial processes to make appropriate decisions. Real-time data and information on the various manufacturing process help the operators to identify the errors, malfunctions, or degradations in the processes.

iii Interoperability: Smart devices, sensor nodes, intelligent machines, and people should be capable of communicating, transferring, and connecting with one another via standard communication protocols.

iv Technical assistance: Assistance systems are technically capable of helping operators with important information. These systems independently perform tasks safely and make decisions independently, with minimum or no human intervention.

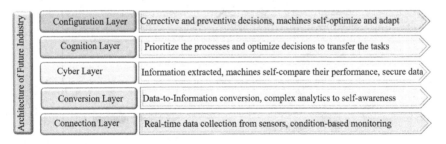

FIGURE 4.7: Future architecture of industries [74].

4.4.2 Industrial plant–The future architecture

The future *Industrial Plant* is the transformation of the existing automation in the traditional industrial processes to autonomous industrial processes. This modification in the industries requires the help of smart and intelligent devices. IIoT connects the industries and other organizations with the help of the Internet. The real-time data generated from the industries are mostly in the unstructured form. Most of the industrial processes are performed locally to avoid delay and preserve the security of data. The data generated in the structured form are transferred over the Internet. Thus, the industrial data are filtered before it is transmitted to the cloud. Fig. 4.8 shows the various layers of a futuristic Industrial Plant. The sensor-enabled devices, logistics, and intelligent machines form the underlying layer of the industrial plant. The advanced analytics performed on the real-time data collected from the sensor nodes, a survey of product quality, and improvement of the quality using feedback are the steps of industrial processes to be performed in a future industrial plant. To perform the computation of time-sensitive tasks and avoid delays, fog nodes act as middleware. The local processing of the data is performed at the fog nodes. As the primary processing is offloaded to the fog nodes, the net energy consumption of the sensor nodes reduce due to requirements for lesser computations. The sensor nodes apply a publish/subscribe approach to communicate with distant nodes [94]. Finally, the processed and analyzed data are provided to the end-users.

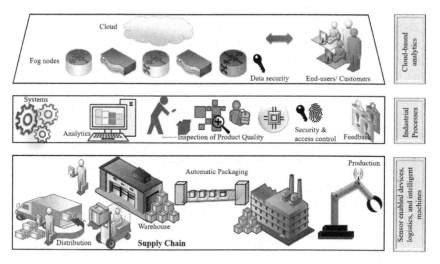

FIGURE 4.8: Block diagram of a future industrial plant [94].

With the recent trends in technological progress, a Cyber-Physical System (CPS)-based five-layer future architecture of an industrial plant was proposed by Lee *et al.* [74]. Generally, CPS has two primary functional components – advanced connectivity and intelligent data management. Fig. 4.8 describes the five-layered architecture and step-by-step workflow of future industries starting from initial data generation, analysis, information extraction, to corrective decisions. The five-layered future architecture of a future industrial plant are as follows:

1. Connection layer: The sensor nodes, machines, and other physical devices are interconnected through the Internet. As outlined in the schematic diagram of the architecture of industries in Fig. 4.7, the real-time data collected from the machines act as the basic structure in the development of future industries. These intelligent devices continuously monitor the autonomous operations of various industrial processes in both normal and hazardous environments, with minimum or no human intervention. Therefore, the safety of workers can be enhanced, and health hazards in industries can be checked.

2. Conversion layer: The predicted second layer of the future industries extracts meaningful information from the collected data. The complex analysis of the data generated from sensor nodes helps to estimate the health of machines, predict faults, and forecast the remaining useful lifetime of the machines.

3. Cyber layer: The primary function of this layer is to extract the additional meaningful information to provide a deeper perception of the performance of individual machines, as shown in Fig. 4.7. This layer forms the main information center of the predicted architecture. The resemblance of trends between the performance of machines and previous historical information helps to predict the future performance of machines. Through extensive analysis, these machines are capable of self-comparing their performance. Further, the security of data is another important function of the cyber layer. The data are necessary to be protected from any form of malicious threats or attacks.

4. Cognition layer: To include the cognition layer in the future industries, detailed information of the entire system is necessary. With the help of a decision support system, integrated simulations, complex analysis, and remote monitoring of the industrial processes are performed. Therefore, the comparative and complete information about the performance of the machine helps to prioritize and optimize the decisions. Combined diagnostics and decision-making are the important functions of this layer, as outlined in Fig. 4.7. To transfer the obtained information to the end-users, complete knowledge of the system is essential. For example, the complex analytics applied to the collected data helps to predict the faults and downtime of machines. Similarly, an automatic photo tagging option on Facebook is an example of the application of analytics. On the other hand, from a business perspective, analytics help to predict the future demand of users, forecast new business opportunities, and improve the product by incorporating the analyzed changes.

5. Configuration layer: This layer provides feedback from the cyber level to the physical or connection layer. The feedback helps the machine to self-optimize and make correct decisions. Based on the decisions taken, the machines self-optimize and self-adapt with the changes in the environment.

4.4.3   Viewpoint of industrial processes

The detailed description and analysis of any specific industrial process is known as the *viewpoint*. The organizations or individuals involved with an organization, who are concerned about the system, propose these viewpoints. Each proposed viewpoint set forth may consist single or multiple views. The collection of ideas of the associated organization/s which describe, analyze, and explain the industrial processes result in the development of viewpoints. From these viewpoints, the architecture of the entire system is designed, and its future state can be predicted. The various viewpoints of industrial processes are categorized as follows:

4.4.3.1   Functional viewpoint

The functional viewpoint primarily represents the functional components of the entire system. It includes the structure and relation among the components and their interaction with external elements. It is the basic point for understanding the functional architecture of a system. The flow of data and control exists within the functional domain. Fig. 4.9(a) illustrates the closed-loop among the data sensed from the sensor nodes, performing analytics, and applying control over actuators. The various functions include the management of systems, configuration, monitoring, and updation of firmware to supervise the operations. The data collected from other layers, mostly from the control domain, a transformation of the data, or analyzing those data improve the intelligence of the system. The application of intelligence helps to execute business functions. The outcome obtained due to control, coordination, and combination of the effects of each functional domain have different levels, and their corresponding cycles vary. As these cycles become longer, their impact increases with the data flow in the higher levels, as shown in Fig. 4.9(a). The extent of data flow becomes larger in the higher levels. Therefore, new intelligence emerges in a larger context with the flow of new information.

4.4.3.2   Operational viewpoint

The operational viewpoint represents the group of functions, which distribute and deploy the resources, manage the resources, monitor and diagnose the system, predict and analyze the outcome, and optimize the systems. This viewpoint unlocks the opportunities facilitated by higher levels. Fig. 4.9(b) shows the block diagrammatic representation of various layers of an operational viewpoint. The principal tasks of operational viewpoint are – predict

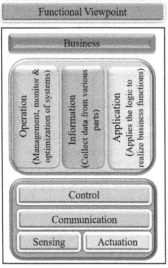

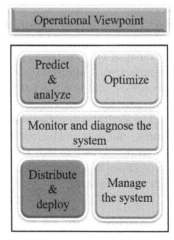

(a) Functional viewpoint.   (b) Operational viewpoint.

FIGURE 4.9: Viewpoint of industrial processes.

& analyze, optimize, monitor, distribute & deploy, and manage the resources. The configuration, registration, tracking, and deployment of the resources are performed remotely and securely. Management of the resources, their control, continuous monitoring, and diagnosis of the system provide an early warning of any abnormal system behavior. Optimization of the system results in the reduction of energy consumption, upgrades the system performance, identifies production loss, and inefficiencies at any level. Therefore, the operational viewpoint provides automation in data collection, processing, validation, recognizing abnormalities, and analyzes the reasons for these problems.

4.4.4 Digital Enterprise

To effectuate the potential of Industry 4.0, Siemens [78] provides solutions for digital enterprises. The benefits of digitalization can be attained in the industrial processes if end-to-end integration of data is achieved. The combination of industrial automation technologies, upgradation of software, communication networks, and security generate the digitally represented value chain.

The digital transformation of the process industries is termed as a digital twin, which helps to optimize the various operational processes and improve their flexibility and efficiency. To maintain the stability of factory units and enable to continue their operations without any obstruction or failure, a digital twin of the industry should be developed. The testbed simulation, testing, and process optimization, before running the processes in the real world, result in the improvement of product quality. Moreover, the upgraded quality of the products is maintained throughout the service life of the plant. Further, the downtime of a plant and their unnecessary maintenance is significantly reduced. The continuous monitoring of processes helps to collect and process data in near real-time, which improves the feedback received to assist in the improvement of the processes.

In manufacturing industries, digital twin assists in the improvement of products, processes, and manufacturing lines. Digital twin helps in the conversion of the product ideas into reality, with the inclusion of the recent technological trends. The digital enterprise primarily focuses on providing products at the lowest price, not only in the automation. The digitalization of the entire value chain is vital to create a digital twin of the machine, production process, and the performance of the machine. Moreover, the entire process includes both suppliers and consumers.

4.4.5   Applications of Industry 4.0

Innovative products, advanced technologies, and various software and development tools have led to the digitalization of systems, in order to enable smart and efficient factory units. To protect the interconnected components, processing and computation of the collected real-time data are performed at the edge of the network. The automation of machines is done using Programmable Logic Controllers (PLC). The smart sensor nodes are connected directly to the PLC system via an ethernet or input-output (I/O) system. Industry 4.0 has a wide range of applications, some of which are discussed as follows:

1. Manufacturing: Industry 4.0 has a significant impact on the domain of manufacturing. There are three types of automation in production, which are as follows: (a) fixed automation,[†] (b) flexible automation,[‡] and (c) programmable automation.[§] Automation in the production line refers to a series of connected workstations. As the production process advances, millions of units are produced, and their real-time statuses updated online. To maintain efficient operation, proper coordination among the parts, activities, and operations is necessary. Automation in the manufacturing industries results in the enhancement of product quality and worker safety. Further, the errors in the production process are reduced, and labor costs are minimized.

2. Logistics: In the Industry 4.0 context, advanced technologies, smart devices, and analytics play an important role in various aspects of supply

---

[†]Fixed automation refers to an automated production facility where the sequence of processing operations is fixed by the specified equipment.

[‡]Flexible automation is an extension of programmable automation. For each batch of a new product, the equipment needs to be reprogrammed.

[§]Programmable automation refers to the automation for production in batches chain management and logistics. The automation in the overall workflow arranges and manages the entire end-to-end process. Automation and new technologies in the supply chain have resulted in the management of various sub-processes such as shipping and warehousing, and inventory management is a centralized process. For example, the automation of the end-to-end process of loading and offloading the mammoth-sized containers into/from ships is a very complicated task. Europe's first robotic port at Rotterdam is fully autonomous, where the trucks and cranes operate autonomously.

3. Agriculture: It took many decades to shift from plough to tractors, and introduce automation techniques in agriculture. Traditional agricultural methods have transformed, and new techniques have been integrated. It is estimated that after two more decades, technological changes in the field of agriculture will be much more significant. Smart and precision agriculture has resulted in an increase in production and improvement in product qualities. With the inclusion of advanced technologies and devices, the laborious and tedious jobs of the farmers have reduced. Autonomous tractors and seed planters help in ploughing, planting seeds at appropriate depths and proper intervals, and applying fertilizers. Additionally, automatic irrigation and maintenance of crops help to improve crop quality. Unmanned Aerial Vehicles (UAVs) or drones can also be used to record infrared, hyperspectral, and ultraviolet images to monitor crop health and analyze soil quality. UAVs are also applicable for spraying fertilizers and liquid pesticides in the agricultural fields.

4. Public Transportation: With the increase in the urban population, the availability of proper public transport has become essential. The automated public transport system primarily aims at reducing pollution and improving the mobility of common people within cities. Driverless metros and driverless buses are some of the developing projects in public transportation. The detection of obstacles, optimization of traffic/vehicle speed, minimum onboard intelligence, and ability to self-monitor are some of the essential functions of an autonomous vehicle. Such autonomous public transport system results in energy savings, provision of cheaper transport, improved user experience, and safety of common people [79]. The appropriate management of door opening and closure at the stops or stations is also a complex task, requiring precision and coordination.

5. Construction: Automation in the field of construction can reform starting from the initial stages of planning and design, operation, maintenance, to recycling the engineering structures. The construction sites are quite dangerous and hazardous. Therefore, the use of advanced technologies and smart sensor nodes improves the safety of the workers. The inclusion of new technologies also enhances the quality of work

with accuracy, the economics of the society, and replaces humans from monotonous and hazardous jobs.

6. Food Production: With the digitalization, the food and beverage industry has witnessed rapid changes. To upgrade the existing machines, identification and modification of system failures, provision of feedback from the data collected in real-time, and tracking of the scrap materials are necessary. Moreover, the safety of machine operators is improved, and the amount of energy consumed reduces. For example, robotic milking of cows and automatic cleaning of the wastes at the dairy farms in the United States has allowed for the scale-up of this industry in a manner that would not have been possible with just human workers.

Summary

This chapter introduces readers with the basic concepts of IIoT, IIC, Industrial Internet Systems, and its impact on the economy, energy consumption, and physical assets. The fundamental of intelligent industrial systems, industrial processes, and architecture of future industrial plants are discussed in this chapter.

(a) The terms Industry 4.0 and Industrial Internet of Things (IIoT) are used interchangeably in academia and industry, but the concepts are not the same.

(b) Industrial Internet Consortium (IIC) is an open membership foundation, founded jointly by AT & T, Cisco, General Electric (GE), IBM, and Intel in 2014 to design new testbeds, share ideas, and insights.

(c) Industrial Internet Systems have resulted in the improvement of income and living standards of common people and workers.

(d) Intelligent machines, connected people, and advanced analytics act as the key factors in the design of Industrial Internet systems.

(e) Smart sensor nodes continuously monitor, control, and provide feedback to improve the various industrial applications.

(f) To include new technologies into industrial processes, distributed decision generation, information clarity, technical assistance of workers, and interoperability among heterogeneous devices are necessary.

(g) In the future industrial plants, sensor-enabled devices would sense and continuously monitor the industrial processes in real-time. Further, advanced analytics help in the improvement of product quality, and filtered data are transmitted to the cloud.

(h) The digital transformation of the industrial processes helps to optimize the running time, improves the flexibility, efficiency of the processes, and product quality.

Exercises

1. Write the differences between IIoT and traditional automation.

2. What is IIC? Write the main objectives of IIC.

3. What are the essential elements of the Industrial Internet?

4. How does Industrial Internet affect global energy consumption?

5. State the different physical assets of (a) steam power, (b) hydroelectric, (c) geothermal, and (d) nuclear power plant.

6. What are smart sensor nodes? Explain the working principle of a smart sensor.

7. Explain the key features of IIoT for industrial processes.

8. Describe the layered architecture of a future industrial plant.

9. What is a digital twin? Discuss the application of digital twins from the perspective of manufacturing industries.

10. Explain the various industrial applications of Industry 4.0.

5

Business Models and Reference Architecture of IIoT

Learning Outcomes

- ■ This chapter covers the preliminary topics related to a business model – the primary units and the requirements of a business model in IoT and IIoT.

- ■ New readers will be able to get insights into various reference architectures of IIoT.

- ■ This chapter will also help the readers to gain knowledge about the Industrial Internet Consortium (IIC) and Industrial Internet Reference Architecture (IIRA).

- ■ The various IIRA frameworks, which are based on different viewpoints, are discussed in this chapter.

- ■ This chapter will describe the basic ideas of the Key Performance Indicators (KPIs) for occupational safety and health of workers.

5.1   Introduction

Business models and reference architectures provide a holistic representation of the overall process in an organization, which involves stakeholders, rules and regulations, vision, capabilities, information, strategies, and the products developed. A business model is the combination of the organizational and financial structure of a business. Theoretical knowledge in economics (business studies) is not a pre-requisite for understanding the concepts of a business model. The core of a business model lies in the process by which value is delivered to customers, attracting customers to buy products, and conversion of payments into profit. The reference architecture of a business provides the basic functional and organizational view of the business. The method which describes the business operations of an organization and information related to businesses is termed as the *reference structure* of the business. It is independent of the functional structure of the organization.

5.1.1 Business models

A *business model* expresses the logic and provides data and additional information, which demonstrates the process of creation of a business and remittance value to the customers. This value is generated from the revenue architecture, costs, and profits related to the business. In order to make a profit from innovation, it is necessary for business executives to work on product innovation and business model design, and understand customer needs. The outcomes of a good business model are value propositions. These outcomes are beneficial for the customers and businesses by attaining advantageous costs and allowing businesses to capture enough value to generate products and services. Customers demand solutions that recognize their specific needs, not just the products. The factors of consumers' selections, transaction costs, and heterogeneity, among the customers and producers, dominantly affect a business model.

5.2 Definition of a business model

A business model describes the rationale of how an organization creates, delivers, and captures value. [80]

5.2.0.1 Primary units of a business model

Fig. 5.1 illustrates some basic units of a business model. The detailed description of the primary units of a business model [80] are as follows:

(a) *Value Proposition*: Value Proposition indicates the selected products or services which fulfill the requirements of a customer. Values provided to the customer may be quantitative or qualitative, as demonstrated in Fig. 5.1. Value is created depending on the originality of products, improvement in services, and adjustment of products as per the customer's requirements. For example – the advent of mobile phones came up as a new industry around mobile telecommunication. Moreover, a value can also be created by assisting customers in their work, through user-friendly designs and optimal selection of product price. The low price value propositions tend to have a significant impact on business models. Service Level

Agreements (SLAs) help to minimize the risks incurred by customers while purchasing products or services. The value of any product can also be improved by making products easily accessible to customers.

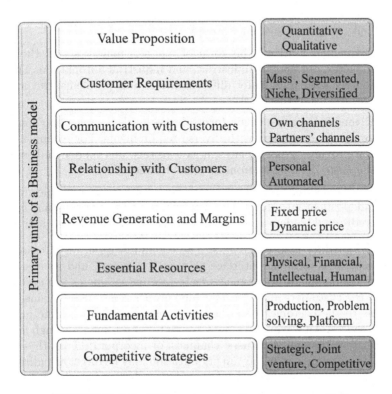

FIGURE 5.1: Different units of a business model.

(b) *Customer Requirements*: Customers form the core of a business model. Based on their requirements, a business model may be described as one or multiple customer groups. A business organization or company solely depends on the profits, which are primarily influenced by customers. In order to serve the customer's requirements appropriately, a proper distinction between the customer's needs to be served and the requirements to be ignored is important. Distinction between these needs is made based on the customer's needs, reachability of the communication channel of customers, the relationship among the customers, and the amount they are willing to pay for products/services. As shown in Fig. 5.1, different types of customer segments are as follows – mass, niche, segmented, diversified, and multi-sided markets.

(a) Mass market: This type of market is available in the consumer electronics section. The customer relationships in these markets focus on a group of consumers with similar needs and problems.

(b) Niche market: This type of market is found in the buyer-supplier sections. The business models that follow niche markets aim at specific and specialized customer groups.

(c) Segmented market: This form of markets mainly serves the watch industry, automobile industries, and industrial automation. The customers with marginally different needs and problems are provided services using this market.

(d) Diversified market: This form of the market serves two unrelated groups of customers with diverse requirements. For example, Amazon provides a cloud platform as a storage and on-demand server to the customers, apart from its retail business.

(e) Multi-sided market: This form of the market serves multiple groups of independent customers. For example, a credit card company consists credit card holders as well as merchants who accept these credit cards.

(c) *Communication with Customers*: The communication, distribution, and sales channels form the interface between a company and its customers. These communication channels create awareness among the customers, assist the customers in evaluating value propositions, deliver value propositions to the customers, and help customers purchase specific products and services. An organization may contact its customers through its communication channels, partners' channels, or incorporate both these channels. The channels owned by an organization may be direct or indirect and yield higher margins. Owned communication channels can be costly in the initial phases. On the other hand, partners' communication channels are always indirect and result in lower margins.

(d) *Relationship with Customers*: The type of relationship maintained by the company with the various group of customers acts as another building block in the business model. As depicted in Fig. 5.1, an organization may establish a personal or automated relationship. The personal relationship is based on human interactions with customers after the completion of purchase. In some organizations, each individual client is served by a dedicated representative. Further, certain organizations do not maintain any direct relationship with the customers, either system generated or similar necessary measures are provided to help the customers. On the other hand, some organizations interact with the customers through the formation of various communities.

(e) *Revenue Generation and Margins*: The revenue generated from each customer group in an organization acts as the brick of a business model. Each of the revenue streams generated may consist of different pricing mechanisms. A business model consists of two type of revenue streams – *transaction* and *recurring* revenues. The transaction revenues are generated from the one-time payment of customers. On the other hand, recurring revenues are generated from the payments delivered from the customers at regular intervals (such as a subscription). The revenue may be generated in various forms, such as selling the assets of the organization, revenue collected from the sale of products or services, subscription fees, money collected from rents/money lent, advertising, intermediate services provided to other parties, and money exchange in the form of licenses. A company generates these revenues in the form of fixed and dynamic prices. Fixed prices are generated from the predefined variables depending on the static variables. Dynamic prices vary based on market conditions.

The margin gained by a company describes the profit earned by the company based on the costs incurred to operate the business model. Value creation and delivery, maintaining relationships with customers, and generating revenue incur costs. Therefore, to enhance the profit gained by a company, the costs incurred should be minimized. The costs incurred in running a business can be curtailed by reducing the cost structure or improving the value of the product. The cost structures of a business model may be fixed or variable. A fixed cost structure refers to the cost structure, which remains fixed or static with the volume of goods or services produced. On the other hand, the variable cost structure refers to the cost structure, which varies proportionally with the volume of goods.

(f) *Essential Resources*: The resources or assets play an important role in a business. These resources enable the organization to create and offer a value proposition, reach the market segment, maintain relationships with the group of customers, and receive money. The assets owned/leased by any organization can be physical, financial, intellectual, or human. Physical assets consist of physical objects such as buildings, vehicles, machines, and systems owned by an organization. Financial assets of an organization consist of financial resources, credits, and cash owned by the organization. Intellectual property includes the patents, copyrights, partnerships, and customer databases possessed by the organization. The human resources maintained by a company are also considered as assets in a business model.

(g) *Fundamental Activities*: To successfully run an organization, a business model must follow certain fundamental activities. These fundamental activities can be categorized into production, problem-solving, and platform. The production activities are related to the design, production, and delivery of better quality products in significant quantities. The timely and satisfactory response to the query of individual customers is also a part of the fundamental activity of an organization. The platform also acts as an essential resource in businesses, which influences the fundamental activities of the organization.

(h) *Competitive Strategies*: The network of suppliers, partners, and consumers forms the business model. To optimize the business models, reduce risks, and build a relationship with customers, companies maintain strategic relationships with competitors. There are four types of distinct partnerships of an organization – strategic, competitive, joint venture, and buyer-supplier relationship, as given in Fig. 5.1. The buyer-supplier relationship is the most commonly designed partnership, which optimally allocates the resources. In most of the organizations, the resources owned by the company or the activities performed are shared with other partners to minimize costs. Further, in a competitive environment, strategic alliances help to reduce risks. Some organizations may perform their activities or share their resources on their own. To improve access to customers through licensing, knowledge, these companies may form a partnership with others.

5.2.0.2   Requirement of business models in IoT

The major challenge for IoT stakeholders is to create value for businesses. Further, the stakeholders may face problems in identifying the needs and opportunities, addressing the internal organizational issues, and various ways to overcome the complexities associated with new IoT technologies. The issues related to IoT-based products are required to be processed and addressed in real-time. IoT ecosystem* consists of a wide range of heterogeneous systems, which provide solutions to various complex industrial processes as given in Fig. 5.2(a). Thus, the existing business frameworks and business models are unable to address the challenges involved with the developments of new products, efficient processes, enhanced asset utilization, services, and technologies. The various aspects to be fulfilled by the IoT-based business models are as follows:

(a) Each business has to provide services to a group of targeted customers, identify, and resolve their problems while designing the model. IoT-based business models should be capable of supporting the customers beyond the organizational level to the ecosystem level.

(b) The values of the customers should be fulfilled. Each company offers a selected group of products and services to meet the requirements of the customers.

(c) The design of complex value streams within the stakeholder network helps to build and distribute the value proposition for the organization. The

---

*A business ecosystem comprises the various values related with the system, which aims at the methods of creating value for both the organization and the entire system.

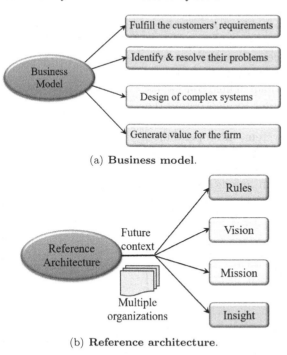

(a) **Business model**.

(b) **Reference architecture**.

2: Requirement of business models and reference architecture.

a new business model includes products and activities, appro-
ources, capabilities, and proper coordination among them.

iness model considers data as an asset, and it is associated with
ess. The primary element of a firm is to generate value from the
cets such as revenue mechanisms and cost structures.

le, Airbnb is an online company that connects customers with
he exchange of a certain amount of processing charges. Airbnb
ketplace and provides real-time information about apartments,
g classes, adventures, surfing, and concerts to its customers. The
ted customers is huge – ones who can access the website globally,
ctive services as per their requirement. On the other hand, the
consumers have proper coordination among them. The business
d by them is built upon economy sharing and generates value.

erence architecture

chitecture provides a standard framework, which explains busi-
functional, and implementation viewpoints. Based on the ref-
ecture, systems are developed and deployed, even considering

ns in the future. In other words, the reference architecture of a busi-
nprises functions and interfaces, which assist in the step-by-step imple-
on of complex technologies. A software reference architecture refers to
' level system architecture, which provides the details of the software
es, elements, relations, and properties among them. Thus, continuous
of the reference architecture is necessary in order to incorporate the
ghts into the system, as shown in Fig. 5.2(b).
apid increase in complexity, the size of the system, interoperability,
nological changes drive the development of the reference architecture
tem. Additionally, reference architecture provides a standard vision,
ny, and context, which promotes communication among the various
ations. The stakeholders of any organization/s actively work and com-
e among themselves during the implementation of any solution. This
reducing errors, and the delays caused.

rence architectures capture the essence of existing architectures and
vision of future needs to provide guidance to assist in developing
system architectures. [83]

3usiness Models of IoT

nects the physical devices or things using sensors, actuators, and var-
nmunication technologies through the Internet. To commercially ex-
e technologies, new business models are developed. A business model
s the framework of the key activities of the business. The various types
ess models [84] of IoT are as follows:

cription-based model: The applications and services provided by the
providers in the IoT scenario are commonly available to the customers
scription, pay-per-use based model. In other words, the services pro-
e in the form of *metered* services. The customers pay only for whatever
have been used by them over a given period, instead of any form of a
e payment. However, the organization must be capable of supporting
scription-based model from the traditional one. This transformation
a significant change in the processes, rules, people involved, and the
nfrastructure. The services provided are acknowledged after the incre-
payment from the customers. The organization realigns the revenue
ed from the customers in the form of *recurring* revenue. In order to
ad collect the *recurring* revenue, new policies, systems, and processes
ired. Moreover, revenue can be earned from the products associated

lel through upgradation of payment methods or implementing
model. Due to the repetitive post-subscription interaction being
the customers, this model has an ongoing relationship with the
wever, the major disadvantage of this model is the management
ers, the requirement of a skilled workforce, and periodic updates.
based model: The business or products deliver the outcome to
in a pay-per-outcome-based model. For example, the end-users
hcare services pay based on the outcome of the service. These
ervices have the capability to self-monitor themselves at regu-
Thus, the customers need not bother about the maintenance
t. The transition occurs from providing a product or service to
s via transactions to directly providing the outcome to the cus-
ected devices help in the formation of outcome-based business.
s pay for the outcome obtained from the products or services,
in a reduction of risks, improvement in profit margins, and in-
ner satisfaction. However, the standardization of the prices, safe
e outcomes, new infrastructures, and policies are the significant
be addressed in the outcome-based model.
ring model: The business equipment, skills, and personnel knowl-
underutilized over time are shared with the customers for a
period. The services are offered in the pay-per-time period ba-
nizations provide access to their equipment and personnel on a
Therefore, these companies earn revenue using a common plat-
mary aim of the organizations undergoing this business model is
he downtime of the equipment and maximize asset utilization.
reate a successful asset sharing – financial, sustainable, and so-
ities are necessary. Further, appropriate policies and rules are
e set, and the supply and demand inventory is required to be
arly. For example, an online platform, named DOZR, was initi-
ry 2017 to provide heavy equipment on a rental basis. Thus, asset
value to the existing resources, enhances profit margins, and re-
Additionally, it reduces ownership over a product and improves
s' access to the resources. The evolution of digital technologies,
tivity, and coordination act as catalysts in driving these busi-
However, it is also essential to address the shift in risks between
d service providers.
service: In an IoT scenario, the physical devices or things are
d through the Internet, which enables businesses to provide the
he consumers on lease. These products can be software, hardware,
and data. The consumers use these products on a pay-per-use
IoT devices, sensors form the core of these connected systems.
e physical and digital worlds co-exist with each other. As the
connected, they are provided as service repeatedly. The major
e addressed in the IoT environment is interoperability. The data
the heterogeneous sensors should be compatible with them to

e productivity. Further, any fault during the run time may lead to in-
data generation. On the other hand, IoT-as-a-service reduces licensing
hances revenue, and improves customer relationships.

› may be *reliability-based* and *reputation-based* business models. The
ty-based model enables consumers to access the IoT products directly.
ing on the product quality and services offered, the price of the prod-
l services fluctuates. Therefore, the supplier/service providers need to
n and improve the quality of products and services at regular inter-
a the other hand, with the improvement in the quality of products
vices, their prices also increase. Therefore, there is an improvement
›rofit of the service provider, as well as customers' satisfaction. The
on-based model is also designed from the consumers' perspective. In
del, based on the history of services successfully delivered by the ser-
vider and within the stipulated time duration, the price of the product

3usiness models of IIoT

hers have put forward business models to meet the requirements of the
mation of the technical potential to economic value. Further, innova-
cepts have been introduced to establish a valuable connection between
urces and the market. The major components of a business model in
› as follows:

le Proposition: Value is created for the product, which follows certain
s or regulations before the actual product or service specifications are
›rmined. From the users' context, the initial point is to identify the
ositions which enhance the value-in-use. On the other hand, from the
ʼ perspective, the value propositions in the case of manufacturers, who
er buy or lease products, can avail transparency, real-time data, and
otely access the data.

le Capturing Mechanism: The mechanism by which a service provider
sforms the present value of assets into ones with financial value is
wn as value capturing mechanism. The primary motivation behind the
gn of business models for IIoT is to widen potential revenue streams
ond the sales manufacturing equipment and services. In other words,
business models evaluate the agreement, which comprises renting,
ing, maintenance and repair, predictive modeling, and process opti-
ation.

le Network: The state of changes in any business organization and
anced connectivity of business is expressed through a value network.

network design of an organization is capable of utilizing its
opportunities to provide services to the customers. Therefore,
network design of the ecosystem of stakeholders and suppliers
anization is necessary. The key to the configuration of the inter-
machines, devices, and processes in industries is networking.

mmunication: The value creation of any organization depends on
tion and interaction among the entities. Moreover, various com-
and uncertainties exist in IIoT. There is a widespread require-
olve the complexity and uncertainty in communication among
tors in a business model. Therefore, communication, trust, and
volved play an important role in business.

iness opportunities in IIoT

which are not known a priori or are unpredictable are referred
nty in economics. IIoT paves the way toward the creation of new
rtunities. A business opportunity comprises sale or lease of any
ce, or equipment required to set up the business. The evolution
logies plays an important role in the new digital economy. Ser-
, system integrators, and the technology partners enjoy mutual
selling the improved product. Thus, the business opportunities
two different aspects – Entrepreneurship and Transaction cost.

eurship aspect: The business models require the ownership of
ssets that create value and are related to the organization. The
ity and workers' experience can be improved through the col-
of humans and machines. Traditionally, the production process
sented as humans versus machines, which has now been replaced
ncept of humans work with machines to deliver new value-added
. The service-based business opportunities primarily focus on the
based on pay-per-use. Depending on the real-time information
in the IIoT scenario, the service providers design the organiza-
ese assets.

on cost aspect: Non-ownership value propositions, such as shar-
hysical devices or things (assets) over a certain period, is an
t aspect of earning revenue. Based on the process of asset shar-
wners manage the downtime of the assets and maintenance costs
ter transparency and control over the process. In the IIoT sce-
h the continuous increase in the number and types of industrial
t, the end-users demand service-level agreements (SLAs) from
e providers.

Categorization of business models in IIoT

y IIoT is widely applicable in the manufacturing as well as non-
:turing industries. Traditional automation took place in the indus-
ring the 1960s when the machines were not interconnected through
rnet. This infrastructure has been drastically transformed to deliver
tion from sensors to the server. The business models in IIoT are cat-
. as follows:

ud-based business model: The customers in the cloud-based busi-
; model do not purchase software, platform, or infrastructure. The
d resources are provided as a lease to the customers for a short time.
he traditional business model, the sensor data is allocated to the ser-
provider via the Internet. Further, the raw data is transformed and
lyzed by personnel. In the IIoT scenario, the outcomes such as pro-
ing power, storage space, virtual operating systems, and development-
nted platforms are offered "as-a-service" to the customers. The ser-
provider focuses on the sales resources, instead of selling the same
luct repetitively to the customers. This business model comprises
:essing power, data storage, and virtualization of the system. In the
astructure-as-a-Service (IaaS) model, the hardware and software are
vided to the customers on a payment basis. On the other hand, in the
: of Platform-as-a-Service (PaaS) business model, the outcome can be
:ssed by external entities, support the development of software appli-
ons, and help in integrating the applications into existing solutions.
her, in the Software-as-a-Service (SaaS)-based business application,
services are offered to the customer as online and customized software
ices.

vice-oriented business model: The service-oriented business model
sts in primary utilization, analysis, and aggregation of data. The cus-
er enjoys the outcome without actually owning it. The transformation
he traditional business models into service-oriented business mod-
results in value creation. These models serve a group of consumers
request services through infrastructures and platforms established
he cloud-based business models. The consumers generally avail ser-
s through a self-service interface or web portal. The type of services
vided to consumers is mostly automatic. On the other hand, the value
figuration of these business models can be achieved through the main-
nce and development of platforms, infrastructures, and applications.
core competency of these business models exists in the fulfillment of
customers' requirements through platforms, data analysis, and real-
e data. The consumers make payments to avail of these services based
he initial establishment structures and variable costs. Further, the rev-
e collected by the service providers may, directly or indirectly, affect
monetary conditions of data.

(c) ***Process-oriented business model***: The process-oriented business model is a hybrid business model, which involves the characteristics of both cloud-based and service-oriented business models. The service provider analyzes and converts the data into a meaningful form. Process optimization reduces downtime and increases the availability of the machine. The optimization of the process proves beneficial to the entire organization. The primary application spheres of the process-oriented business model are 3D printers, process-oriented, and data-oriented structures. The key features of the process-oriented business model are–execution of various business processes within the organization, presence of process owners to supervise the processes, and evaluation of process performance. With the help of data collected and its analysis, the process flow can be optimized. The initial establishment cost of this type of business model is quite high.

The cloud-based business model primarily focuses on providing infrastructure. The companies using a service-oriented business model employ cloud for data storage, analysis, and providing the outcome as-a-service. On the other hand, process-oriented business models utilize both cloud- and service-oriented business models to serve their purpose. The major challenges faced in the development of IIoT-based business models are:

(a) As the physical worlds and virtual worlds combine at a large scale, it is quite difficult to preserve the security and handle data privacy issues.

(b) The organizations require improved security frameworks, which provide authentication at the device-level and system-wide assurance, rigidity, and incidence response models.

(c) With the lack of interoperability among the existing systems, the complexity and deployment cost of the systems increase significantly.

(d) To share data seamlessly between the machines and other physical systems (M2M communication) will be quite expensive for the addition or replacement of the existing technologies.

5.5   Reference architecture of IoT

Internet of Things (IoT) interconnects millions of physical devices or things in the form of a network, with the help of the Internet. Each of these devices possesses a unique identifier and consists of in-built sensor nodes. These sensor nodes provide real-time information about the environment, which provides beneficial insights from the sensed data. As shown in Fig. 5.3, the sensed data

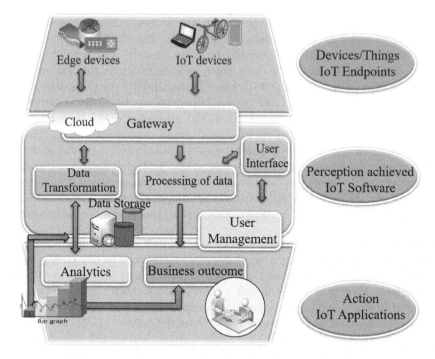

FIGURE 5.3: Reference architecture of IoT [81].

from the sensors are transmitted to the cloud through a gateway. The time-sensitive data are primarily processed at the edge devices. The sensed data are processed, transformed into meaningful form, and stored in the cloud. Further, analysis of these data helps to provide a particular business outcome. The IoT applications are used to track physical objects; medical applications assist in remotely monitoring patients, connected vehicles, and remotely monitor various household applications. Moreover, the IoT platform is used to deal with risks related to the network, manage the data, and help to maintain data security. The smart IoT processes are automated using complex analytics, security at the cloud-end, and utilize advanced technologies. The primary principles of a reference architecture for IoT are as follows:

(a) *Heterogeneity*: The framework should be capable of supporting various types of devices, environments, processing patterns, scenarios, and standards. The architecture also assists different hardware and software heterogeneity.

(b) *Security*: To develop a secure system, it is necessary to establish a connection between the physical and digital world.

(c) *Large scale deployment*: The reference architecture is deployed on a large scale, such that it supports millions of connected things.

(d) *Adaptability*: As the framework is deployed on a large scale it supports heterogeneous devices. With the combination of several building blocks, a solution is designed to serve the IoT infrastructure. The extension points help to integrate the existing systems and applications.

Based on the type of applications, the reference architecture of IoT can be classified as *Stateless stream processing* and *Stateful stream processing*, as shown in Fig. 5.4. Stateless stream processing applications are straightforward and easy to build compared to stateful stream processing applications. These applications do not store data. Generally, stateless applications have only one function or service. Depending upon the information received with each request, the server processes them without relying on the previous state information. The stateless applications can also be represented as a pack of microservices. From the resource point of view, the arrangement of these stateless applications determines the best location for the usage of the system. Reading the news from the Internet and connecting it through HTTP is an example of stateless processing.

Stateful stream processing applications can be referred to as a database. Unlike stateless applications, stateful stream processing applications process the requests based on the information received from earlier requests. The same server processes the requests associated with the same state information. In other words, all the same state information is required to be shared with the servers who need it. From the application point of view, the arrangement of these stateful applications determines the best location for the usage of the system. In the stateful applications, the state information between the client requests is maintained by acknowledging the transaction's existence and the session's existence between the requests. Additionally, the server, which processes these requests, and workload management service, which consider both the similarity of server and transaction, are considered in stateful applications.

5.6 Reference Architecture of IIoT

Industrial Internet Reference Architecture (IIRA) identifies and emphasizes on the architectural patterns of IIoT systems in the industrial sectors. Additionally, IIRA categorizes the architectural patterns depending on the viewpoints of the stakeholders. The simplified and summarized view of the IIoT infrastructure may be represented in various forms.

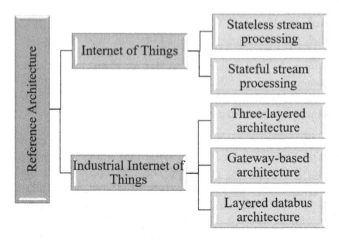

FIGURE 5.4: Types of reference architecture.

5.6.1   Categorization of reference architecture in IIoT

The abstracted view of an IIoT subsystem can be represented in the form of different structures, as shown in Fig. 5.4. The standard IIoT reference architectures are as follows:

(a) *Three-tier IIoT architecture*: The three-tier IIoT architecture comprises three layers – edge, platform, and enterprise layer. As shown in Fig. 5.5, data flow and control flow play an essential role in processing the activities of each layer. The data from the sensors and actuators, involved in the various complex industrial processes, are transmitted to the edge devices within their communication range. Based on the specific use-cases, the edge devices include their processing area, location, and nature of the devices. The sensors, actuators, devices, and control systems connect with an edge node or multiple edge nodes present within their proximity network.

The platform layer receives data from the edge devices, primarily processes those data, and assists in communicating commands from the enterprise layer to the edge layer. Data flow from both the edge layer and the enterprise layer to the platform layer. Further, the platform layer analyzes and manages the data. Connectivity is established with the other two layers through the Internet or over cellular 4G/5G network. The enterprise layer receives data from both the edge layer and the platform layer. The other two layers receive commands from the enterprise layer. The enterprise layer acts as an interface to the end-users and other industry personnel associated with the system. Further, this layer connects and communicates with the other two layers via a private network or the Internet. Therefore, the edge layer executes the primary processing,

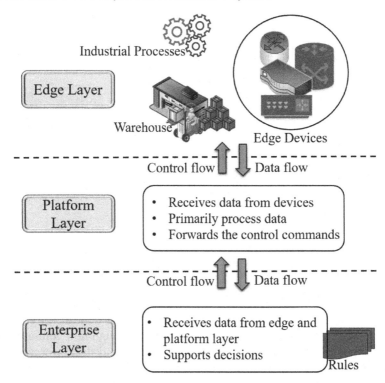

FIGURE 5.5: Three-layered IIoT architecture [82].

platform layer works on the operational aspects, and the enterprise layer is concerned about the business aspects and applications.

(b) *Gateway-based IIoT architecture*: The gateway-based IIoT infrastructure provides local connectivity among the sensor nodes and actuators to the wide-area network through the gateway. As shown in Fig. 5.6, the gateway acts as the mediator between the sensor nodes and the wide-area network. The local area network comprises the connected heterogeneous sensor nodes and transmits their data to the edge layer, as illustrated in Fig. 5.6. In other words, the gateway layer isolates the local area network from the wide-area network. Generally, the edge network uses hub-and-spoke (star) and mesh topology. In the hub-and-spoke topology, the edge gateway acts as the hub, and the group of edge nodes are the spokes connected to the hub. In the mesh topology, the edge gateway also acts as the hub, and the other edge nodes are connected through the wide-area network. However, in both hub-and-spoke and mesh topology, the wide-area network cannot access the edge nodes directly.

As the primary processing and analysis are performed at the edge layer, the complexity of the IIoT systems is significantly reduced. Moreover, the

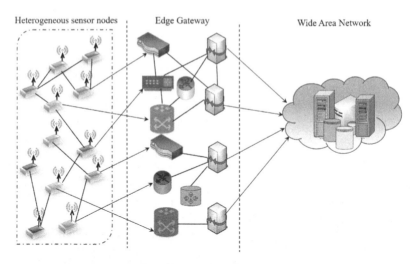

FIGURE 5.6: Gateway-based IIoT architecture [82].

management of the resources, devices, and assets is done at the edge layer. The edge nodes help in local communication through new communication technologies and protocols, local data transfer, and data processing. The local data processing at the edge nodes includes several functionalities such as data aggregation, transformation, filtering, and analysis. The data is transferred from the edge nodes to the wide-area network through the edge gateway via wired serial buses and short-range wireless networks. Further, the gateway manages the edge nodes locally and the data remotely. The location decisions and application logic are generated based on the local edge nodes.

(c) *Layered Databus IIoT architecture*: The layered databus-based IIoT architecture is typical in various industries. Across the different layers of this architecture, the latency reduces, and secure and peer-to-peer data communication is facilitated. The direct communication between the various applications is useful in the field of control, local monitoring, and edge analytics. Fig. 5.7 shows the schematic diagram of the layered databus architecture, where smart machines form the base of the architecture.

Smart machines help in local control, automation, and real-time analytics using databus. Further, at the system level, the monitoring and supervisory control are performed. The system-of-systems level deals with more complex, Internet-based networks, controls, complex analytics, and monitoring. The logically connected space between a set of schema, which communicates using those schemas, is known as a database. Each layer of the database helps in achieving interoperability between the endpoints. Additionally, the databus supports communication between the system-level

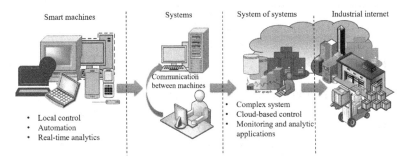

FIGURE 5.7: Layered databus IIoT architecture [82].

applications and heterogeneous devices. At the system-of-systems level, the databus connects multiple systems simultaneously. The communication within the architecture in the databus occurs through a *publish-subscribe* communication model. The various advantages of the layered IIoT infrastructure are–quick device-to-device integration within milliseconds or microseconds, automated data delivery of applications within the buses, isolation of hierarchical subsystems, and integration of sensors and actuators.

5.7 IIRA

Industrial Internet Consortium (IIC), an open membership organization, has published the Industrial Internet Connectivity Framework (IICF) to unlock the data in isolated systems. IIC has also announced a common framework, system, and application architectures for IIoT termed as IIRA. IIRA is a standard, open infrastructure, which helps in interoperability, maps the applicable technologies, and helps in all-round development. Fig. 5.8 illustrates the reference architecture of IIoT. As shown in Fig. 5.8, the stakeholders put forward their viewpoints and identify their concerns, which are then incorporated during the development of models. Finally, the viewpoints of stakeholders and their feedback are considered. The developed system model is improved based on the feedback, and the final system model is generated.

In the industries, IIRA establishes and highlights the critical concerns of the architecture. Various concerns are classified based on the viewpoints of stakeholders. Some of the essential terms related to the reference architecture are discussed below:

Architecture viewpoint: The specific information related to any infrastructure and analysis of the system is termed as the viewpoint of the architecture.

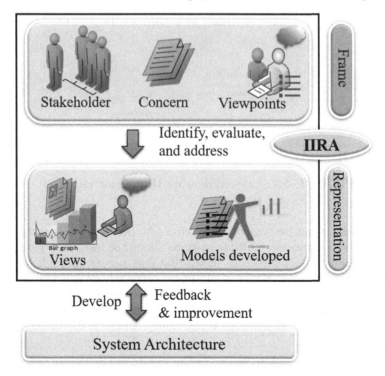

FIGURE 5.8: Industrial Internet reference architecture (IIRA).

Stakeholder: Any individual entity or organization with common interests related to the system or infrastructure viewpoint is considered as a stakeholder.
Architecture view: The accumulation of various ideas, which describe, analyze, and provide a solution to the specific concerns in a viewpoint, is referred to as the architectural view.
Architecture framework: The combination of the ideas of stakeholders with similar interests and their representation lead to the development of the architecture framework.
The detailed description, analysis, and representation of the IIoT reference architecture are general and provide an abstracted view related to the diverse industrial applications. IIRAs are continuously refined and revised based on the feedback provided from the real-world deployed IIC testbeds of the IIoT systems.

5.7.1 IIRA framework: Basics

Different stakeholders may have varying interests related to their specific systems. In order to build and evaluate a system, it is necessary to identify and classify the views of the stakeholders appropriately such that it reveals the full life cycle of the system. To fulfill this requirement, IIC proposed the Industrial Internet Architecture Framework (IIAF). IIAFs consists viewpoints, principles, and practices for the proper description of the IIoT architecture. These standard frameworks help in more straightforward implementation and effective decision-making of the architecture viewpoints proposed by the stakeholders. Further, these frameworks act as a resource, which facilitates the development, documentation, and communication of the IIRA. The system developers can use these frameworks to describe, develop, and organize the representation of the system.

5.7.2   Categorization of IIRA frameworks

IIRA frameworks are designed and developed in the footsteps of the views put forward by the stakeholders. Based on the use cases developed by IIC, IIRA has four different viewpoints–business, usage, functional, and implementation viewpoints [82]. The four viewpoints are as follows:

(a) *Business viewpoint*: In a business organization, stakeholders play an essential role in determining future directions.

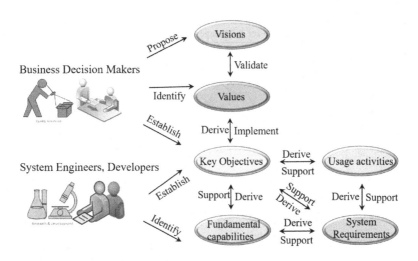

FIGURE 5.9: Value and vision-based business model [82].

Fig. 5.9 illustrates the detailed flow diagram of the visions, values, key objectives, and fundamental capabilities of an organization from the business perspective. The stakeholders are the *decision makers* in an organization, who are automatically incorporated in the implementation of the new system. Further, values indicate the vision recognized by the stakeholders.

The future state of any organization is illustrated by vision. In the context of the value, key objectives of a business organization and the deliverable outcomes of the business are determined. On the other hand, the high-level specifications of the system to complete any specific business tasks are referred to as its fundamental capability. The primary base to identify the fundamental capabilities in businesses are independently described. From the business viewpoint, the stakeholders identify the vision of the organization and enhance the system operation, through the adoption of IIoT. After that, the system developers suggest values and organize them to implement and improve the system. The objectives put forward by the stakeholders help to derive the key objectives and fundamental capabilities of the organization.

(b) Usage viewpoint: The human or logical users denote a sequence of activities to be addressed in the user viewpoint. This viewpoint represents the expected usage of the system. Further, the usage viewpoint also illustrates the activities, which synchronize with the different units of the system, and accordingly, execute the tasks. Fig. 5.10 depicts the workflow diagram of the usage viewpoint model.

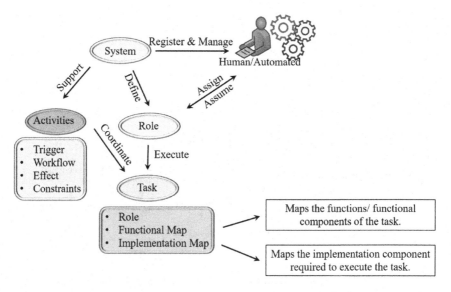

FIGURE 5.10: Usage viewpoint model [82].

The activities of an organization act as necessary inputs for the system requirements, manage the system design, implement, deploy, and help in the evolution of the IIoT system. A particular task is executed considering the role decided by an entity. The role represents the set of capacities assumed by an organization to initiate, execute, or participate in the task. Any entity or organization performs one/multiple roles in order to perform a task. Each of the tasks performed has three parts – role, implementation, and functional map. The functional map describes the functions or functional components, which map the task. On the other hand, the implementation map describes the implementation components involved in the task execution. Any activity performed by an organization is a detailed description of the coordination among the tasks/activities. The description of the task organization is necessary to define the step-by-step procedure of an IIoT system. The same activity may be repeated multiple times if required. Each activity is composed of a trigger, workflow, effect, and constraints. The primary condition/s, which initiate the execution of the activity is/are termed as the *trigger*. In order to perform an activity, the tasks are organized sequentially, parallelly, conditionally, and iteratively. Any effect illustrates the differences between the state of the system after the activity is performed. Further, the system characteristics are obtained during the execution and formation of the new state.

(c) Functional viewpoint: Industrial Control Systems (ICS) utilize control systems for the management of different processes and machines. ICS helps in the improvement of overall productivity by reducing the cost of production and minimizes the error rates. Further, the combination of operational and information technology results in the upgradation of the production line safety of the workers.

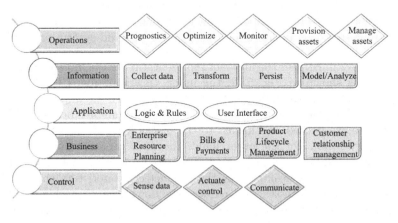

FIGURE 5.11: Functional viewpoint model [82].

Fig. 5.11 shows the five important functional viewpoints in the IIoT system. Typically, in an IIoT system, the functional domains are classified as operation, information, application, business, and control. The control domain represents the group of various functions, which help in the sensing, actuation, control, and communication among the ICSs. Based on the data sensed by the sensor nodes, control signals provide actuation. The sensors, actuators, and gateways connect to the communication module. On the other hand, the various functions that provision, manage, optimize, and monitor the systems in the control domain are illustrated in the operation domain. Traditionally, ICSs emphasize the optimization of assets. However, Industrial Internet optimizes and manages the type of assets and customers. Based on the historical data, the predictive analysis of IIoT systems helps in the identification of probable issues. The various data from the control domain are collected, transformed, modeled, and analyzed in the information domain. The analytics involved are either done through *online* or *offline* modes. In the information domain, the data consist of functions for syntactic and semantic transformation, data storage, data distribution, and quality of the data. In the case of online streaming, data processing supports real-time analytics. On the other hand, in offline streaming, the data are processed in batches. The application domain comprises logic and application interfaces to understand the various business functions. Further, the business domain helps in the operations of the IIoT-based systems by combining the existing functions with the Industrial Internet-based functions. Connectivity among the components of the system is necessary for the management of the huge amount of data generated.

(d) Implementation viewpoint: The general architecture and technical representation of the overall system, as advised by usage and functional viewpoints, are described by the implementation viewpoint. The technical representation includes interfaces, protocols, and system properties.

Fig. 5.12 illustrates the abstract representation of the implementation viewpoint. As shown in Fig. 5.12, the usage viewpoint maps to the functional components, and the functional components map to the implementation components. For example, in an automatic packaging and product delivery system, products are packed and loaded to the trucks for delivery with minimum or no intervention. On the other hand, the data generated from the sensors, deployed at different locations, are processed and analyzed. The outcome of this analysis helps in improving the product characteristics, enhances the efficiency of the processes, and reduces the downtime of machines. Therefore, the mapping of the entire system characteristics can be performed using the implementation map.

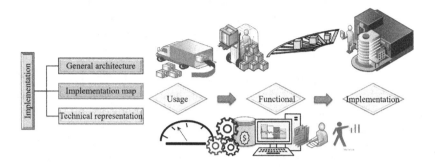

FIGURE 5.12: Implementation viewpoint model [82].

5.8 Key Performance Indicators for Occupational Safety and Health

The performance measure of the overall activities of an organization is termed as its KPIs. The indicators demonstrate the statistical values, which measure the current conditions. Additionally, these indicators also provide future trends and outcomes. KPIs communicate with the customers and provide valuable feedback to improve the organization. In other words, KPIs estimate the effectiveness of the organization toward attaining its business goals. Therefore, KPIs help to access the progress of the organization over a specified period. With the help of specific objectives, measurable progress, achievable, relevant, and time-bounded goals, good KPIs are designed [85]. The continual evaluation and re-evaluation of KPIs result in assessing the performance of the organizational activities. KPIs are categorized into leading and lagging KPI based on the outcome indicated by them.

(a) *Leading KPIs*: Leading KPIs are mostly used to predict the economics of an organization. Leading KPIs are hard to measure, easily controlled, and input-oriented. In other words, leading KPIs are the predictive measures of the organizational activities during the execution of a process. For example, the percentage of managers and workers with the appropriate OSH training, number of audits, and frequency of observed unsafe behavior of workers. Therefore, the leading KPIs can be improved with time, which influences on the lagging KPIs.

(b) *Lagging KPIs*: Lagging KPIs indicate the changes after the economy has started. Lagging KPIs are easy to measure, output-oriented, and hard to improve. For example, the number of lost time incident frequency rate, number of production rate lost due to absence caused by sickness, incidents, and number of fatalities. Therefore, lagging KPIs essentially influence the future progress of the organization.

We consider the KPIs for a healthcare organization. The leading KPIs of a hospital are the time taken to provide healthcare services to a patient, turnaround time for different medical tests in the laboratory, and the number of experienced doctors in each of the units. On the other hand, the lagging KPIs include the number of patients in emergency, the number of beds occupied in the emergency in the present scenario, average stay of a patient, and recovery rate of critical patients in the hospital.

Summary

This chapter introduces the business models and reference architectures of IIoT. The business model reflects the organizational and financial structures of an organization. On the other hand, reference architectures characterize various operations and information related to the businesses. Further, the necessity of the business model and reference architecture in IIoT is also discussed.

(a) In order to commercially explore the technologies, different business models are developed in IoT, such as subscription-based model, outcome-based model, asset-sharing model, and IoT-as-a-service.

(b) The major components of a business model in IIoT are value proposition, value capturing mechanism, value network, and value communication.

(c) The evolution of new technologies in IIoT has resulted in the setting of new business opportunities. IIoT has two different aspects – Entrepreneurship and Transaction.

(d) Various business models of IIoT are classified as cloud-based, service-oriented, and process-oriented business models.

(e) Lack of interoperability, data privacy, and security issues, and authentication of data at the organizational level are some of the major challenges of IIoT-based business models.

(f) In the IoT scenario, sensor nodes sense and transmit data to the cloud/edge node, depending on the time-criticality of data. Further analysis of the data provides a particular business outcome.

(g) Based on the type of applications, the reference architecture of IoT is classified as Stateless and Stateful stream processing.

(h) The reference architectures of IIoT are categorized as three-tier IIoT architecture, gateway-based IIoT architecture, and layered databus IIoT architecture.

(i) IIC, an open membership forum, announced the common framework, system, and application architecture for IIoT, termed as IIRA.

(j) IIRA framework is designed and developed based on the business, usage, functional, and implementation viewpoints.

(k) KPIs measure the performance of the overall activities of the organization. Based on the outcome, KPIs are classified as Leading or Lagging.

Exercises

1. Define a business model. What are the various units of a business model?

2. Give an application market scenario for segmented and multi-sided markets.

3. Define a reference architecture. Discuss the principles of the reference architecture in IoT scenario.

4. Discuss the challenges to be experienced in IIoT-based business models.

5. What are the differences between stateless and stateful stream processing?

6. Discuss the data flow in a three-layered IIoT architecture.

7. What is IIRA? Discuss the basic architecture of IIRA.

8. How are IIRA frameworks classified?

9. (a) Write the differences between business and usage viewpoint.
   (b) Write the differences between business and functional viewpoint.
   (c) Write the differences between implementation and usage viewpoint.

10. What are the KPIs of an organization? Discuss the importance of KPIs in terms of business and reference architecture of an organization.

11. Write the differences between leading and lagging KPIs.

12. Give some examples of leading and lagging KPIs for (a) healthcare and (b) manufacturing industries.

Technological Aspects of Industry 4.0 and IIoT

6

Key Technologies: Off-site Technologies

Learning Outcomes

■ This chapter provides basic knowledge of the key aspects of cloud computing and fog computing in the industries.

■ New readers will get an overview of the necessity of cloud and fog in the various industrial sectors, cloud platform providers, and Service Level Agreement (SLA).

■ With respect to the industrial applications, service-based, and deployment-based cloud models are discussed in this chapter.

■ Based on the industrial cloud platform and services provided, industrial cloud companies are briefly discussed in this chapter.

■ The necessity of cloud computing, advantages of fog-enabled industries, and its applications are also reviewed in this chapter.

6.1 Introduction

Advanced technologies such as machine-to-machine (M2M) communication, big data technologies, smart sensors, and complex analytics, form the main platform for the development of Industry 4.0 and IIoT. To transform the traditional industries to smart factories, these advanced technologies assist in provisioning sensing, communication, computing, and networking. Additionally, these technologies enhance the scalability and performance of the entire system. With the integration of these advanced technologies, machines and industrial processes can be remotely operated by various levels of employees and workers in industries. Further, the combination of operational and information technologies help to achieve digitization in the manufacturing and production processes. This results in increased productivity, improved product quality, and enhanced efficiency, which helps in attaining cheaper and durable products [87].

Real-time data is sensed by smart sensors, which are further processed and analyzed with the help of machine learning at the edge devices or cloud. Further, these data provide meaningful insights into the processes and helps automate tasks. Complex and hazardous tasks are performed with minimum or no human intervention. On the other hand, a colossal amount of data is generated from the interconnected devices. Thus, proper handling and storage of these heterogeneous (in terms of volume, variety, velocity, veracity, and value) and unorganized data is a complex task. In order to store, process, and analyze the data, the cloud is required. Cloud computing acts as one of the driving forces toward the development of information technology and is necessary for the successful adoption of Industry 4.0 and IIoT across various industries. There is a wide range of applications of cloud computing; however, integration of cloud with physical processes and intelligent devices is a complicated task. This chapter discusses the necessity of cloud computing in the context of IIoT, various models of cloud, service-level agreement (SLA), and fog computing and its applications in IIoT.

6.2   Cloud Computing

Cloud computing relies on the provisioning of on-demand cloud services such as networking, storage, software, database, processing, computation, and analytics, to the customers. Cloud services support both scientific and business processes. With the help of the cloud, businesses are capable of configuring, designing, and providing customized services to the users. Additionally, the cloud supports a wide range of mobile devices, which helps to access data securely and remotely. Cloud computing has a diverse range of applications across several industries, such as automobile, manufacturing, healthcare, entertainment, retail, and banking.

6.2.1   Necessity of cloud computing

In the Industry 4.0 and IIoT scenario, multiple heterogeneous devices are interconnected. These connected, intelligent devices produce a huge volume of data, which are mostly generated in an unstructured format. The generated data have volume, velocity, veracity, and variability. On the other hand, the processing, storage, and real-time analysis of these data is a complex task. Therefore, cloud computing is necessary for provisioning on-demand services. The purpose of the cloud platform in the industries is shown in Fig. 6.1.

(a) Millions of connected devices generate almost zettabytes of data at every

FIGURE 6.1: Necessity of cloud.

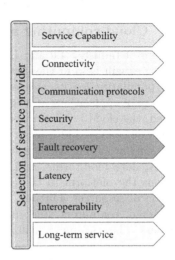

FIGURE 6.2: Features of CSP.

time instant in the IIoT scenario. Cloud computing supports the high computational speeds necessary for processing, storage, and complex analysis of this huge volume of data.

(b) Scalability is one of the essential characteristics of the cloud, which assists in provisioning resources dynamically to the customers on-demand. Additionally, the resources are adjusted, upgraded, and new service requests are automatically adjusted.

(c) The sensor data are vulnerable and are prone to threats. Any data tampering or manipulation of the data may endanger the entire system. Cloud services are secured through a set of policies, rules, and applications. Therefore, cloud services provide a trustworthy and reliable system.

(d) SLA binds both the customers and the service providers in terms of quality of service (QoS), implementation, and service availability.

(e) Cloud services provide an efficient data acquisition system for the collection and processing of the data over the network.

(f) Consumers can easily utilize cloud services and pay according to the resources consumed by them.

(g) Lastly, integration of cloud services help in minimizing the overall business cost of storage, process, and provide meaningful insights from the data.

6.2.2   Cloud computing and IIoT

In the era of Industry 4.0 and IIoT, the integration of cloud with advanced technologies will help in the development of business. Cloud computing has gained popularity for its indigenous features and the ability to support a wide range of applications among the industrial sectors. Based on the services provided, the cloud is categorized as Software-as-a-Service (SaaS), Platform-as-a-Service (PaaS), and Infrastructure-as-a-Service (IaaS). SaaS provides access to software applications to customers via a network. For example, data analytics, data monitoring, and decision-making. On the other hand, PaaS provisions the platform to the consumers for the creation, installation, and execution of resources. Further, IaaS delivers the infrastructure such as virtual storage, computers, and infrastructure components as services to the consumers. The consumers utilize these resources and pay as per their usage duration or amount of resources used. Based on the deployment type, the cloud is categorized as – public, private, and hybrid. Therefore, cloud computing enables ubiquitous, on-demand services to the end-users [91]. Here, the end-users are the industrial sectors, who avail services from the service providers. The services requested by the various industrial sectors may vary. The different end-users of cloud in IIoT and Industry 4.0 may belong to the domain of healthcare, transportation, supply chain, manufacturing plants, oil refineries, agriculture, mining, or others.

The cloud service provider is the entity, who manages the entire infrastructure, runs the software to provide services, and delivers the services to the end-users. The cloud platform provided by the service provider to the consumers enables them to develop and run the applications. This platform collects data from the sensors, processes, and analyzes the data to provide meaningful decisions. Therefore, the selection of the appropriate cloud service provider is important as the consumers utilize the platform provided by them for the development of various applications [92]. Fig. 6.2 shows the different criteria to be checked by consumers before the selection of a service provider.

- In the Industry 4.0 and IIoT scenario, millions of devices are interconnected through the same platform. Further, with the increase in the number of connected devices, the complexity of handling them increases proportionally. Therefore, the service provider must be capable of connecting a single customer to single/multiple data centers.

- Connectivity is one of the important aspects of Industry 4.0 and IIoT. Thus, low latency and low packet drop are desirable to establish proper connectivity.

- Currently, MQTT and HTTP are the most common and popular communication protocols used for communication. Therefore, the platform provided by the cloud service provider should support these protocols.

- The platform provided by the service provider should be capable of measuring the latency, detecting abnormalities, and taking corrective actions.

- The cloud platform should provision authentication and end-to-end data encryption, and provide certificates of authenticity.

- In case of any fault or error in the network, the cloud platform should be able to recover and take back-up of the data.

- Interoperability is one of the important aspects of IIoT and Industry 4.0. The cloud platform provided to the consumer should be able to integrate heterogeneous data.

- In case of time-critical data, the cloud platform should be capable of processing the data at the edge layer, with minimum or no latency.

- The industrial cloud platform provider must be capable of providing services for extended periods.

- The cloud platform deployed by the service provider should comprise the features of both public and private clouds.

6.2.2.1   Cloud computing services in IIoT

From the Industry 4.0 and IIoT perspective, cloud computing service models are categorized as – SaaS, PaaS, and IaaS. The service models of the cloud for provisioning services in industries [90] are as follows:

(a) SaaS: The cloud service provider deploys, maintains, and upgrades the applications, as per the consumers' requests. The consumers access these industrial applications through a web interface. The consumers enjoy these services on a payment basis. On the other hand, the back-end is managed by the service provider. For example, the industrial machinery catalyst developed by Siemens is a SaaS platform used for industrial purposes.

(b) PaaS: The cloud service provider controls and manages the infrastructure for the platform. Additionally, the cloud service provider also supports the development, deployment, and management of the processes running in the cloud. PaaS service provides end-users the platform for the development of the industrial applications. The clients have control over these applications. However, the clients do not possess any control over the infrastructure. For example, Predix from General Electric, Sentience from Honeywell, and Mindsphere from Siemens are some of the examples of PaaS providers for industrial applications. Additionally, software development firms such as Cumulocity, Bosch IoT, and Carriots provide PaaS platforms.

(c) IaaS: The cloud service provider possesses the physical resources such as servers, networks, and storage. The end-users utilize the computing resources available from the network. Therefore, end-users have more control over the resources. IaaS is used to deploy PaaS and SaaS. For example— Microsoft Azure, Google compute engine, IBM smart cloud, and Amazon web services, are some of the examples of IaaS.

6.2.2.2 Deployment-based models of Cloud

Based on the deployment models, the cloud infrastructure can be categorized as follows:

(a) Public cloud: The cloud infrastructure and computing resources are available over the network for application in the industry. The virtualized resources are publicly shared with the consumers. The cloud service provider deals with the maintenance of the infrastructure, on-demand resources, and servers. Example: Amazon Web Service, Google Compute Engine, and Microsoft Azure.

(b) Private cloud: The private cloud set-up infrastructure is for a single organization, such that the virtualized resources are shared directly with the clients. Private clouds are managed by the clients themselves. In terms of security, private clouds are more secure than public clouds.

(c) Hybrid cloud: The cloud infrastructure which is set up with the unique features of public and private cloud is known as a hybrid cloud. The hybrid cloud platform is flexible and allows the transfer of data and applications between the public and private clouds.

6.2.3   Industrial cloud platform providers

Industrial cloud platforms primarily focus on operational technologies. To improve manufacturing processes, industrial cloud platform providers collect data in real-time with the help of smart sensors. Post data collection, processing, and analysis of these data help to predict process failures. Based on the prediction, corrective measures are taken to correct faults and resume normal operations. This significantly reduces downtimes in industries, thereby resulting in savings of time, money, and precious resources. The three industrial cloud companies which use industrial cloud platforms are – General Electric Predix, Siemens Mind Sphere, and Honeywell. On the other hand, some software development firms using cloud platforms are – C3IoT, Uptake, and Meshify.

- *General Electric Predix*: With the help of the Predix platform, industrial organizations are capable of transforming the way they monitor, manage, and optimize their assets and operations. Predix provides platform-as-a-service to build and run applications. This helps to improve pro-

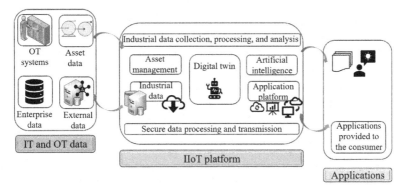

FIGURE 6.3: General electric cloud platform [95].

ductivity, reduce downtimes, and increase the efficiency of the processes [95]. The schematic diagram of the General Electric Predix cloud platform is represented in Fig. 6.3. The Predix platform is unique and is designed to serve certain industrial features, which are as follows:

— This platform supports huge volume, variety, and velocity of data generated by the industries.

— It ensures the security of data and processes, which satisfy the industrial standards.

— Latency and bandwidth requirements are fulfilled for real-time response.

— The industrial control systems are combined with external environment variables.

The main focus of the GE platform is to collect and analyze data generated from industrial machines and their environment, to improve the efficiency of the machines. Predix platform also supports distributed computing for modeling and analysis of assets. Further, the Predix platform enables differentiated functions of applications, which are expressed as a structure of subsystems, sub-assemblies, and components of an asset. On the other hand, the digital twins of various assets can be developed on the Predix platform to understand, predict, and optimize these assets. To remotely operate the various industrial processes, which are located in harsh environments, Predix integrates two complementary software stacks – edge and cloud computing.

• *Siemens MindSphere*: MindSphere is a cloud-based, open IoT platform, which securely connects machines, systems, and plants to the digital world. Industrialists expect the zero downtime of machines with the advent of digitization. Therefore, to minimize downtime of assets, prior identification of faults, and the adoption of corrective actions are necessary. Further, the machines and equipment connect to the industrial cloud. MindSphere collects the data generated from the devices and performs complex analytics to transform the data into productive businesses [96]. The meaningful insights obtained from the data helps to increase the productivity and efficiency of the businesses. The main features of Siemens MindSphere are shown in Fig. 6.4.

- The hardware and software solutions are securely connected with the assets.
- The platform can be easily deployed, implemented, and tested with previously configured solutions.
- Data can be visualized and explored with the help of the cloud platform.
- Data analysis helps to provide meaningful insights.

- *Honeywell*: Honeywell provisions secure IIoT solutions across various industries. It supports various SaaS-based business models. With the help of Honeywell, the existing business models can be transformed through innovation, technology, and experience. The cloud platform of Honeywell is secure, scalable, and standardized. Further, it is capable of providing digital twin of the processes and critical assets, which has the capability of indicating overall improvements in operations [97]. Honeywell is an efficient solution provider for the oil and gas industries. The various key features of Honeywell are shown in Fig. 6.5, which are as follows:

  - It establishes secure collaboration across the entire system in real-time.
  - The analysis of the real-time data provides information and helps in making decisions.
  - The connected assets, processes, and devices help to improve the safety, productivity, and profit of the organization.
  - Data management and control helps to optimize the performance of the organization.

- *C3IoT*: C3IoT offers services to analyze and predict based on the real-time data collected from various industrial sectors. The C3IoT platform utilizes AI and big data applications to deliver predictive insights from the data [98]. The various products of C3IoT are illustrated in Fig. 6.6.

  - C3 AI Suite: This is an integrated software that enables data integration by reducing the complexities involved with AI-enabled applications. The various types of applications of C3 AI Suite are predictive maintenance, inventory optimization, energy management, sensor health, and fraud detection.

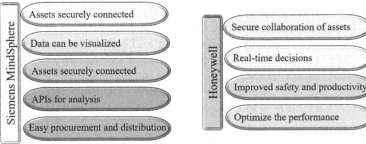

FIGURE 6.4: Features of Siemens [96].

FIGURE 6.5: Features of Honeywell [97].

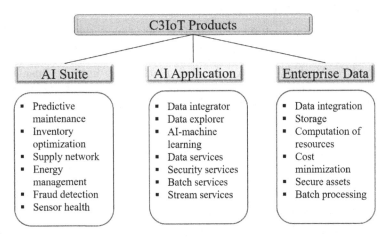

FIGURE 6.6: Products of C3 IoT [98].

- C3 AI Appliance: This product helps organizations that perform analysis of the huge volume of data. C3IoT platform is designed to operate and scale a private cloud flexibly. This product is fully integrated to provide robust development and test the enterprise-scale AI applications (Fig. 6.6).

- C3 Enterprise Data Lake: This product provides a platform for the integration, governance, and analysis of the large volume of heterogeneous data. This product allows access to the data through RESTful APIs.

• *Uptake*: Uptake provides innovative solutions to the customers, which optimizes cost, increases availability, enhances customer satisfaction, and continuously improves the performance of the organization. The complex heterogeneous form of data is transformed into real-world outcomes. Additionally, Uptake has business conversations that serve to identify the goals and strengths of the organization, resulting in achieving better outcomes more smartly. With the help of AI techniques, new business opportunities are created and operational workflow improves [100]. The various industrial solutions of Uptake are – agriculture, road, construction, manufacturing, fleet, energy, equipment dealers, oil and gas, rail, and mining. The key features of Uptake are discussed in Fig. 6.7, which are as follows:

- Improved utilization of resources through the effective computation of operational costs, avoiding replications, and an overall increase in production.

- Continuous flow of revenue, simplicity in the process of buying or selling, and constant flow of revenue.

- Improving customer satisfaction, securing services, smart storage solutions, and maintenance of legal rules and regulations.

- *Meshify*: Meshify provides the IIoT platform to improve business outcomes through digital transformation. It efficiently converts existing devices to Internet-connected products. In this way, the organizations are able to monitor their product in real-time [99]. Meshify is capable of providing these applications at a low cost. Fig. 6.8 illustrates the different products of meshify, which are discussed as follows:

  - Now: This product can be connected and deployed within minutes. Now collects data from various smart sensors and common industrial protocols. The end-users receive alerts through email and text.

  - Tracker: Tracker is a next-generation technology that provides a solution for tracking assets and is compatible with LoRaWAN. This product provides real-time status and activity of processes using advanced techniques such as satellite images or maps. The expenses of using this product are low compared to RFID or Wi-Fi tracking.

  - Enterprise: This is a complex platform, which provides advanced solutions for full-stack IoT development. Enterprise offers customized design and notifies users through emails and messages. Further, this assists in time-series data retrieval and storage.

6.2.4   SLA for IIoT

A SLA is an agreement or mutual understanding between the end-users and service providers, assuring that the minimum level of service quality will be maintained. In the IIoT and Industry 4.0 environment, most of the applications run in real-time, for which SLAs ensure the safety and security of individuals and devices. Therefore, the SLA must be able to attain the goals as per requirement. The SLA metrics may vary from one service provider to

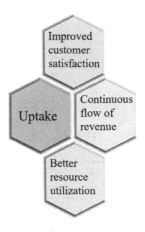

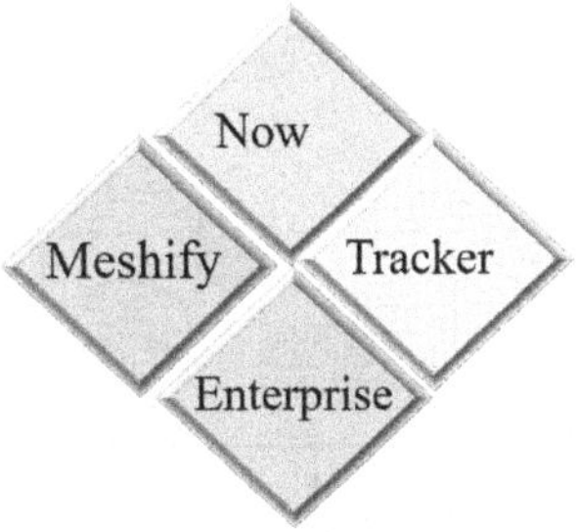

FIGURE 6.8: Products of Meshify [99].

FIGURE 6.7: Features of Uptake [100].

another [88]. However, the precision, accuracy, responsiveness, and availability should be almost uniform. The SLA helps cloud service providers in assuring customers regarding the quality of deliverables. From the client's perspective, SLA helps them to keep a check on the reputation and QoS provided by the cloud service provider. The various characteristics of a good SLA are:

- The service provider must be able to provide the level of service, as mentioned in the SLA.

- The SLA must be relevant to all the customers as given in the agreement.

- The consumer should be capable of measuring the actual level of service and compare with the terms given in SLA.

- SLA should be affordable for the consumers, and it should not impact other services.

- The consumers and service providers should mutually agree with the terms given in the SLA.

- The service provider must be capable of exercising control over the factors, which determine the service level.

The terms of an SLA, which satisfy with IIoT and Industry 4.0 standards, are difficult to design. The management of SLA in the cloud and industries is essential. There are certain reasons such as standardization of SLAs, QoS provided, life cycle management of an SLA, and lack of SLA policies, which makes the services provided by one service provider dependent on the services of others. Thus, standardization of the SLA is necessary. Further, the frameworks for IIoT is not yet properly developed or standardized. The specific

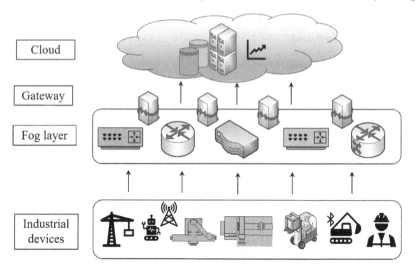

FIGURE 6.9: Architecture of fog layer.

definition of SLAs is necessary for the development of other services in industries. Additionally, there are insufficient SLA rules and regulations to abide by the service provider and the consumers [89].

6.2.5   Requirements of Industry 4.0 and its solution

The transformation of the traditional factories to smart factories and the integration of advanced technologies result in increased productivity, improved quality of the product, and increased efficiency. Industry 4.0 and IIoT aims at provisioning connectivity among the devices, machines, systems, and products. These systems can be remotely monitored and controlled by on-site personnel. On the other hand, real-time data collected from the environment assists in the improvement of the performance of the system. Sophisticated analytics is utilized to predict abnormalities in the system, which helps to upgrade the lifetime and reduce downtime of machines.

The use of a decentralized or distributed approach with the cloud may result in the fulfillment of the requirements of Industry 4.0. The data sensed and transmitted by sensor nodes may be time-sensitive. Due to the decentralized nature, the data which are time-critical is locally processed, and immediate response is provided to them. Any form of undesired delays in processing time-sensitive data such as data related to the health of machines, fire hazards, natural disasters, and accidents may result in life-threatening situations. Therefore, fog computing emerges as a reliable solution to process these time-critical data and minimize latency.

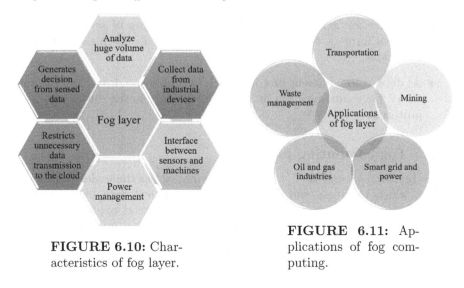

FIGURE 6.10: Characteristics of fog layer.

FIGURE 6.11: Applications of fog computing.

6.3 Fog Computing

Fog computing is a virtualized layer between the connected devices/machines and the cloud. It provides computation, storage, and networking services. However, the fog layer is not a replacement for the cloud layer. The fog layer facilitates computation and processing of time-sensitive data [93]. Intelligent devices that are deployed at the edge of the network are known as fog nodes. Heterogeneous fog nodes are deployed in a distributed manner at various geographical locations. These fog nodes efficiently operate in real-time to provide various services such as filtering, aggregation, and translation. As the time-critical data are processed at the fog nodes, the amount of data transmitted to the cloud is reduced. The fog nodes primarily process these data and, after that, transmits the data to the cloud. The basic architecture of fog is illustrated in Fig. 6.9. Industrial devices and machines sense and transmit the time-sensitive data to the fog layer. The fog nodes transmit these data to the cloud through the gateways. Fog computing serves a wide range of applications such as transportation, smart grid, and the healthcare system.

In the IIoT and Industry 4.0 environment, data is sensed and transmitted continuously by the smart sensor nodes, which are deployed at various industrial locations. These data vary in volume, velocity, and veracity. Additionally, some of the industrial processes require their data to be processed faster and need instantaneous action due to other delays and security measures. The major challenges in handling this huge volume of data are:

- Data is produced from different sources in various forms, which are structured, unstructured, or semi-structured.

- The data are generated from heterogeneous types of sensor nodes in different formats, such as simple byte and complex schema-based representations.

- As data utilize different protocols such as simple broadcast, MQTT, and CoAP for communication, it is difficult to include the diverse types of protocols.

6.3.1   Fog computing for IIoT

In the Industry 4.0 and IIoT, fog computing may act as the middleware technology, which processes and analyzes the time-critical data locally within a specified period. Further, depending on the industrial requirements, the fog layer can be regarded as a cloudlet, an edge layer, a microdata center, or a nano-data center. Traditionally, the fog layer provisions quick delivery of processed data and facilitates dynamic computation offloading of data among the fog nodes. However, in the smart factory environment, the fog layer additionally assists in the transparency of information extracted from collected data, decentralized decision-making, information security, interoperability, and analysis of data [94]. The fog layer monitors the energy consumption of sensor nodes and modifies the amount of data sensed by them. Fig. 6.10 illustrates the characteristics of the fog layer in the Industry 4.0 and IIoT environment, which are as follows:

- Fog layer analyzes a huge volume of real-time data generated from the industrial devices and machines.

- The fog layer simultaneously collects data from various interconnected industrial machines, smart sensors, devices, and robots.

- Fog layer generates the decisions from the sensed data and delivers them to the actuators with minimum latency.

- Fog layer acts as an interface between the sensors and machines, using certain protocols.

- Power management of the entire system is done in the fog layer.

- Fog layer helps to avoid the unnecessary data transmission to the cloud and server by restructuring and filtering the data.

6.3.1.1   Necessity of fog computing in industrial analytics

The introduction of the fog layer as middleware in industries results in enhanced performance, increased efficiency, addressal of data security and privacy concerns, and diminishing operational costs of factory units. Fog computing provides advanced methods for optimized decision-making and performing intelligent operations. To process and analyze time-critical data, analytics and other applications run at the edge of the network so that the data is processed near the source. Additionally, local processing of data leads to autonomous execution of the processes, rules, and algorithms. The uptime of machines improves due to the removal of the round trip delay to the data center. Further, as data is processed and analyzed near the source, it is less likely to be exposed to privacy and security threats. Analysis and filtering of data at the cloud add bandwidth costs combined with the costs of data transmission to the cloud. For example, in a thermal power plant, different processes, such as the conversion of water to high-pressure steam, conversion of steam to mechanical energy, and mechanical energy to electrical energy in the turbine coupled with the generator generate terabytes of data. The transmission of this huge volume of data to the cloud is possible, but it becomes costly on a regular basis. The classification of analytics to process and analyze the data collected from different sensors and industrial processes can be done based on the location and functions performed. The data communication, processing, storage, and maintenance at the cloud data centers are costly. Therefore, the integration of the fog layer to the industries results in upgraded performance, improved privacy and security issues due to local processing, and minimization of operational costs in the long run.

6.3.1.2 Advantages of fog-enabled industries

The computational resources and applications are distributed along the path of data transmission in the decentralized fog computing environment [101]. As the data is computed at the network edge, there are some advantages of fog-enabled industries, which are as follows:

(a) The real-time streaming, analytics, and visualization of the data generated from various industrial sectors at the edge of the network help to reduce the cost of data transmission to the cloud.

(b) The distributed architecture of fog assists in provisioning end-to-end security of the collected data, information, or applications.

(c) Due to the distributed architecture, fog computing increases the flexibility and scalability of the network.

(d) The overall cost of data transmission, processing, and analysis is reduced with the introduction of the fog layer.

(e) Innovative business models and software have resulted due to fog computing architecture.

6.3.2 Applications of fog and their solutions

Fog computing has a wide range of applications across the various industrial sectors, as shown in Fig. 6.11. The different industrial applications of fog include mining, transportation, smart grid, oil and gas, waste management, and advertisement industries. Some specific applications of fog in different industrial sectors are as follows:

- *Transportation*: Transportation is one of the principal constituents of industries. Intelligent Transportation System (ITS) and Advanced Driver Assistance Systems (ADAS) are the important technologies that assist in the improvement of traffic congestion and reduces accidents. Safety of the driver and vehicle form the major part of the road transport industries. In a connected vehicle environment, sensor nodes are either deployed into the vehicles or vehicles possess inbuilt sensor nodes, which sense and transmit data to the cloud or edge layer. Based on the processed information, the safe speed of vehicles, safe distance between the vehicles, and inter-vehicular communication may result in reducing the number of accidents. Further, prior information on road conditions such as the presence of potholes, manholes, and speed breakers, may enhance road safety. The local processing of the time-critical data helps in the delivery of information within the specified time. Internet of Vehicle (IoV)* environment supports in-vehicle communication, smart parking, and smart traffic lights.

- *Mining*: Mining is one of the most hazardous and risky occupations. With the help of smart sensor nodes and sensor-related technologies, the productivity, underground communication, and safety of working personnel have improved. Additionally, the failure of the machines, emission of hazardous gases in the underground mines, and chemical reactions can be predicted a priori from the collected data. Further, the maintenance of machines and devices, and energy efficiency are the important constraints in mining industries, which are significantly improved with the incorporation of fog computing. Therefore, the accuracy of the system is improved using sensor nodes, localized processing, and analysis of data in the fog architecture.

- *Smart grid and power*: Smart grid is an electrical framework that provides two-way communication between the consumers and suppliers. With the help of smart meters, smart appliances, and energy-efficient resources, the smart grid efficiently delivers electricity to the consumers. Further, the smart grid is environment-friendly, reliable, improves the efficiency of the power services, and minimizes the production losses. Smart grid helps in demand-side management through load management, improved utilization of resources, reduction of load during peak pricing hours, and usage of renewable resources. Additionally, demand

---

*IoV is the network dynamically formed with the integration of smart sensors and technologies for smooth handling of traffic flow, fleet management, and avoid accidents.

response support enables the generators and loads to interact in real-time to avoid the sudden rise or fall in demand. The elimination of spikes in demand results in the reduction of wear and tear in machines and devices, increases energy efficiency, and reduces voltage spikes and flickers. On the other hand, an advanced metering infrastructure (AMI) provides real-time information on the energy consumed. AMI comprises smart meters, advanced meters, computational infrastructure, and data management systems. Through AMI and bi-directional communication, data communication and data flow management are improved, and power outages and blackouts are minimized [94]. Further, the forecast of future demands and the amount of power consumed by the load are performed using advanced methods.

- *Oil and gas industries*: Smart sensor nodes and advanced computing/processing technologies are integrated to perform real-time operations. Traditionally, sensor reading anomalies were remotely transmitted to the cloud through satellite links. As the transmitted data was vulnerable to threats and round-trip latencies were high, the introduction of fog computing, reduced the incidents of data breaches. For example, in an oil pipeline, pressure and flow sensors, and control valves and pumps are present. The local processing of the collected data may reduce the round-trip latency. Fog nodes are based on wired supervisory control and data acquisition system (SCADA), interfaces, and Modbus, which help to connect with the industrial devices and machines. Improvement in safety, security, latency, and efficiency of the system are expected with the proper integration and control between the fog and cloud [102].

- *Waste management*: Waste management is one of the important goals in the industries. To transform a traditional industry into smart factories, the management of waste collection and disposal is an essential task. One of the principal objectives of the lean manufacturing approach is to eliminate waste products. The proper management of wastes results in the improvement of product quality, minimization of the product cost, decrease in material costs, and reduction of lead time. Depending on the type and quantity, waste collection and disposal are linked with the recycling process, such that proper resources are allocated. The introduction of a fog layer in this task helps in reducing network bandwidth usage and brings down costs of regular computation and storage at the cloud.

- *Advertisement*: Advertising is one of the important methods to promote new products and services. For example, the promotion of products helps to provide information about new products in supermarkets and shopping malls to the customers. The fog nodes deployed near the network edge communicates with the store's Wi-Fi and transmit promotional information to the customers, depending on their interests and shopping profiles. The customers register with the store, which stores their shopping history. Based on the previous data of the customer, fog nodes can track the interests of the customer and provide them with information on new products, offers, and discounts.

Summary

This chapter introduces the necessity of cloud computing and fog computing in industries. Further, depending upon cloud computing service models and deployment-based cloud models, we discuss the various industrial cloud platform providers. Additionally, we discuss the requirement and applications of fog layer in the various industrial sectors.

(a) Cloud computing provisions on-demand services to the customers. In the IIoT and Industry 4.0 scenario, various industrial sectors are customers.

(b) Based on the services provided, cloud computing can be categorized as SaaS, PaaS, and IaaS.

(c) Cloud service provider is the entity that manages the entire infrastructure and provides services to the customers.

(d) Industrial cloud platform providers such as C3IoT, Uptake, and Meshify, provides software development services. General Electric Predix, Siemens Mind Sphere, and Honeywell provide platform as a service to the customers.

(e) A SLA is an agreement between the cloud service provider and the end-users.

(f) A good SLA mentions the level of service the service provider will be able to provide. SLA should be affordable and relevant to all the customers.

(g) In order to minimize the latency of processing and computation of time-critical data, distributed architecture along with cloud – fog computing, was introduced in the industries.

(h) In the smart factory environment, the fog layer assists in facilitating transparency of information extracted from the data, decentralized decision-making, interoperability, and data security.

(i) Fog layer acts as the interface between the sensors and machines, collects data from the interconnected machines, assist in reducing unnecessary data transmission, and analyzes the data.

(j) Fog computing is widely applicable across different industries such as transportation, mining, smart grid and power, oil and gas industries, waste management, and advertising.

Exercises

1. Why is the cloud platform required in various industrial sectors?

2. What is the main criterion to select a CSP?

3. Give examples of public CSPs.

4. Which software development firms utilize cloud platforms? Discuss the various products of any of these firms.

5. What are the different key features of Honeywell? Which type of business model is supported by them?

6. How Siemens MindSphere minimizes the downtime of machines?

7. Describe the cloud platform developed by General Electric Predix. Discuss their unique features.

8. Write the various industrial solutions of Uptake.

9. What are the various products of Meshify? Discuss briefly each of the products.

10. What do you understand by the term SLA?

11. How is a good SLA characterized?

12. What are the general characteristics of the fog layer in industries?

13. Discuss the basic architecture of the fog layer in industries.

14. Why is fog computing necessary for industrial analytics?

15. Discuss in brief the application of fog layer in (a) transportation industry and (b) smart grid.

7

Key Technologies: On-site Technologies

Learning Outcomes

- This chapter provides a fundamental view of the advanced technologies required for the transformation of factories through IIoT for Industry4.0 standards.

- The reasons for the transformation of traditional industries to Industry 4.0 and the necessary advancements required in the factories are outlined.

- In the case of new readers, this chapter will provide a preliminary understanding of various technologies such as AR, VR, smart factories, and lean manufacturing systems.

- Additionally, new learners will get an idea of big data, its characteristics, storage, and data analytics.

- The characteristics of smart factories and technologies used for the development of smart factories are also briefly discussed.

7.1 Introduction

Industry 4.0 or the fourth industrial revolution, has brought a technological wave in the manufacturing industries. The evolution and development of a large number of tools and advanced technologies through Industry 4.0 will result in the widespread usage of IoT in the industrial sectors. IIoT is the adoption of IoT-based technologies in the industries, incorporating machine learning (ML), and big data technologies. The primary key to the evolution of these new technologies is the access to real-time data and system feedback based on the analysis of data. One of the relatively recent technological developments in the manufacturing sector is the lean manufacturing system, which helps in the elimination of wastes from the processes. A customer-based lean production system maximizes the overall productivity. In contrast, other manufacturing approaches also consider tasks and production perspectives. Self-driving/autonomous vehicles are one such example of technological evolution. These types of vehicles share data with the other neighboring vehicles within the network. Further, the use of advanced analytics helps to interpret, analyze, predict failures, and optimize complex industrial processes. This results in the improvement of the processes and products, and reduce the downtime of machines. Additionally, the safety and security of the working personnel also improve as a result of these approaches.

7.1.1 Need for Industry 4.0

Industry 4.0 or the fourth industrial revolution has led to the development of advanced technologies such as smart factories, big data, virtual reality (VR), and augmented reality (AR) [103]. The fourth industrial revolution promotes the adaptation of a sustainable environment and improvement of the working condition of people in the industries. Typically, the maximum working age of present-day factory workers is quite high, which requires safe, easy to handle, and reliable technologies in these industries. Further, the key contributing factors of occupational safety and health (OSH) need to be improved by providing training to the existing personnel. As the traditional production mechanisms are combined with the new technologies, customization of the overall system becomes necessary.

7.1.2 Transformations required

The combination of digital technologies with the new business models aim to bring the necessary transformations required for the development of Industry 4.0 and IIoT across various industrial sectors. The digitization results in a connected environment, which reduces costs and improves efficiency. For example, the connection from the manufacturing industry to the retail sector, mobility services, smart logistics, supply chain, and smart inventory management system enables proper coordination among these separate entities and optimizes the tasks and responsibilities of each of them. The various necessary transformations for industrial development are as follows:

- Advancements required: The traditional industrial processes need to be transformed for enabling the collection of real-time data, monitoring, and control. As the machines and equipment are the assets of industries, predictive maintenance, remote access to the processes, and modular manufacturing help in the upgradation and long-term upkeep of these assets. The primary component required for this transformation in industrial processes is the sensor nodes. Additionally, as these sensor nodes are energy-constrained, the development of energy-efficient systems is a prerequisite for IIoT and Industry 4.0. Further, management of human resources, proper training of personnel, connected and smart supply chain, and product quality improvement is also necessary for the overall change. To fulfill the customers' requirements and synchronize traditional processes with the new technologies, the companies may modify their existing products and bring new products to the market.

- Business strategies: The specific plans or steps followed by an organization to reach a goal are known as business strategies. The strategies for the digital transformation of any organization should reflect the steps to rapidly progress from ideas to design through prototyping of the product to the launch of the newly designed product in the market. The views put forth by the executives' proposed visions and values derived from their viewpoints contribute to the definition of the new models. Further, the incorporation of customers' feedback results in continuous improvement.

- Human skills: To become familiar with the transformations done in the industries, people require adequate technical skills. In the IIoT and Industry 4.0 scenario, a huge amount of data is generated, which needs to be processed, analyzed, and stored for future reference. Further, creative thinking, leadership capabilities, decision-making, and coordination with the others in the organization are necessary qualities to be developed among the personnel of the industries.

- Changes toward value creation: The business strategies followed by the industrial personnel to adapt and incorporate various upgradations to the traditional infrastructure lead to value creation in the industries. To incorporate the technological changes in the factory floor, workplace safety and security of the workers should also be taken into consideration.

7.2   Augmented Reality

AR describes the direct or indirect, reformed version of the physical, real-world environment, which provides superimposed computer-generated effects to enhance the end-users' view [104]. Alternatively, AR combines real-world surroundings and computer-generated effects to provide an amplified view of the physical world. These advanced AR technologies present an interactive and digitally manipulated environment to the users by adding digital elements to the actual environment to produce an enhanced view. As shown in Fig. 7.4(a), the main hardware components used for the AR technologies are processing, display, sensing, and projection unit. Currently, AR technologies are in use in a vast range of industries such as manufacturing, healthcare, food and beverage, fashion, education, marketing, and retail. The AR technologies are widely used to impart safety training to personnel in various industries, especially the ones dealing with hazardous materials and working environments [105]. Further, to optimize processes in the primary stages, AR technologies are utilized to simulate and design the processes, a priori. Consequently, AR plays an important role and is widely applicable in Industry 4.0 and IIoT. The key features of AR are as follows:

(a) AR creates the perception of mixed reality spectrum.*

(b) AR produces enhanced view of the real world using multiple sensory modalities – auditory, haptic,[†] somatosensory,[‡] olfactory,[§] and visual.

(c) The utilization of multiple sensory modalities creates an immersive effect of the real environment.

(d) AR utilizes the existing physical environment and overlays additional useful information on top of it.

7.2.1   History of AR

The concept of 'augmented' view of the physical world was first initiated in the year 1901 [105]. L. Frank Baum, an author, introduced the idea of electronic display/spectacles. During the year 1957-1962, Morton Heilig, a cinematographer, created an idea of a simulator named *Sensorama*, to produce visual, sound, vibration, and smell effects. Further, the first wearable computer with text and graphics was created by Steve Mann. To study the celestial bodies and actual sky images through a telescope, Douglas George and Robert Morris designed a prototype in the year 1987. The designed astronomical telescope-based *heads-up-display* (HUD) is popularly referred to as the predecessor of AR. Thomas P. Caudell, a Boeing researcher, coined the term Augmented Reality. However, the first functioning AR system, named *Virtual Fixtures*, was designed in the US Air Force Research Laboratory by Louis Rosenberg in the year 1992. Further, in the year 1999, US Naval Research Laboratory developed a prototype, termed as *Battlefield Augmented Reality System* (BARS), to provide awareness and training using the wearable system. The *Google Glass*

---

*Mixed reality implies the environment created by the combination of the physical and digital world.
[†]Haptic refers to the perception achieved as a part of the effect to be capable of active exploration of surfaces through touch.
[‡]Somatosensory refers to the creation of enhanced effect produced as a part of the complex sensory nervous system.
[§]Olfactory refers to the perception created to produce the enhanced view with the help of smell.

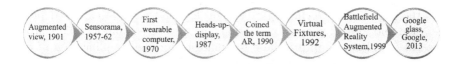

FIGURE 7.1: History of AR.

launched in the year 2013 connected to wireless services on a user's cellphone to provide information on the glass's visor. Later, popular games and applications such as *Pokemon Go* and *Digital Lightfield* were introduced 2016 and 2017, respectively. The various stages of the development of AR are shown in Fig. 7.1.

7.2.2 Categorization of AR

AR technologies are categorized into various types, each with varying differences in technologies and application-based use cases. The different types of AR technologies, as shown in Fig. 7.2(a), are as follows:

(a) *Marker-based AR*: Marker-based AR technologies utilize a camera and some visual marker (typically head-mounted display unit) to produce an outcome. When the camera and visual marker sense a reader, an output is produced. The main function of the camera is to differentiate between the physical objects and a marker. The output from the camera comprises various content/information. Further, the output helps in the computation of the position and orientation of the marker. These type of applications are simple, easily recognizable, and require low processing power.

(b) *Markerless AR*: Markerless AR technologies utilize various devices, which are embedded with GPS, digital compass, velocity meter, or accelerometer, to track the data-based location of the device. Markerless AR technology is also known as position-based or location-based AR. This technology can be used for a real-time, vision-based 3D tracking system [107]. The different sensors used for vision-based 3D tracking are GPS, gyroscope, camera, and accelerometer. Markerless AR technologies are cheaper, robust, and accurate [106]. The 3D model-based tracking depends on the similarity between the local features in the image and the model to be developed. In order to track the objects locally, a moving edge tracker is used. The virtual objects can be projected on the screen using this technology.

(c) *Projection-based AR*: Artificial light is projected into real-world surfaces in the projection-based AR. These AR technologies allow human interaction with the projected artificial light. The computer images/graphics are projected onto a physical 3D model to provide a realistic effect on the object. This technology distinguishes between the expected projection and altered projection. Various manufacturers and suppliers around the world are adopting these technologies for the simplification of complex processes such as manufacturing, assembly, sequencing, and training of operators [108]. A digital operating canvas is virtually created on the work surface in front of the operator. The real-time data available from these processes help in adjusting the assembly lines, improving the quality of the product, and enhancing efficiencies. The user has the capability to touch and hold the physical form of the computer-generated graphics, with their bare hands.

(d) *Superimposition-based AR*: Superimposition-based AR technologies utilize a partial or complete original view of the object with the newly augmented view of the object. Object recognition plays an important role in these AR technologies. This is widely used in medical, education, and military training. For example, an AR-based app can provide a virtual idea of the placement of the furniture in real environments, simply by downloading the selected pages from the product catalog.

7.2.3   Applications of AR

There are wide range of applications of AR technologies, as shown in Fig. 7.3(a), which are discussed as follows:

1. Military: The HUD unit assist in various AR applications in the field of military training. During flight simulation, this device displays the necessary data to the pilot at the aircraft's instrumentation. The fighter pilot has a transparent display of the various data such as altitude, airspeed, and the horizon line [112]. On the other hand, to locate the enemy within their Line-of-Sight (LoS), HMDs are utilized by ground troops. These technologies are also used for training purposes of the troops in the simulator. Virtual maps and 360° field of view helps the soldiers to navigate in unknown terrains.

2. Supply Chain and Logistics: The various applications of AR helps to increase the efficiency of processes at a cheaper rate in different sectors such as warehousing, transportation, and optimization of routes, associated with supply chains.

3. Healthcare: AR technologies are widely used in the field of medical science to perform complex surgeries in the domain of Neurosurgery, MRI, and X-Ray systems. For medical training purposes, students use AR headsets to examine various parts of the human body. In the case of critical patients, real-time information on the exact location of the origin of abnormality in the patient's vitals helps in diagnosis, surgery,

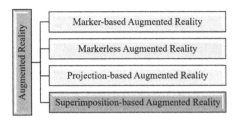

(a) Augmented Reality (AR).

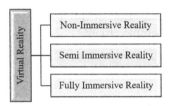

(b) Virtual Reality (VR).

FIGURE 7.2: Types of AR and VR.

and treatment [113]. Devices such as AccuVein can scan the vein of a patient, and MRI or X-Ray systems reduce the risk of the operation.

4. Entertainment: There are various AR applications which are applicable for sightseeing as well as gaming purposes. In order to provide a realistic view of the different historical displays in a museum, tourists can use a smartphone with a camera, which can overlay information about the items/relics being viewed. On the other hand, games such as Pokemon Go, Zombies, and Temple Run are some of the recent advances in AR technologies in the gaming sector.

5. Education: With the upgradation of various technologies, smartphones and tablets are presently being used for education in schools and colleges. To get the reformed version, text, graphics, video, and audio are superimposed to enhance the real-time learning environment of students. These technologies assist students to participate and interact with the environment. For example, AR technologies can help students to learn and visualize the molecular structures of various chemical compounds,

(a) Applications of AR.   (b) Applications of VR.

FIGURE 7.3: Applications of AR and VR.

witness the 3D visualization of the solar system, constellations, and be guided on building simple circuits.

7.3 Virtual Reality

The concept of VR is derived from the concept of a virtual environment and the reality, which human beings experience in the real world. VR generates a realistic 3D artificial environment, which is a mixture of software and hardware. HMDs along with headphones are used to provide a *fully immersive* experience. Fully immersive experience means the user is completely immersed in the non-physical (artificial) environment and naturally communicates with the artificial environment in such a way as if the user is present in the real world. Summarily, VR provides both mental and physical engagement of the user in the virtual environment. The senses of users are stimulated to create an illusion of the virtual world, and after that, actions are performed by them in this virtual world. The various components of VR are input devices such as a mouse, joystick, tracking balls, HMD, and console or personal computer, as shown in Fig. 7.4(b). In the Industry 4.0 and IIoT scenario, VR-based technologies are widely used by users to gain real-world experience. The key features of VR are as follows:

(a) VR creates and enhances computer-generated graphics.

(b) VR provides the user with the perception of being physically present in a non-physical world.

(c) Both auditory and visual sensory feedback are incorporated to give a fully-immersive experience to the users.

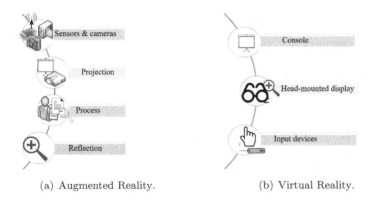

(a) Augmented Reality.   (b) Virtual Reality.

FIGURE 7.4: Components of AR and VR.

FIGURE 7.5: History of VR.

7.3.1 History of VR

Morton Heilig [109] proposed the concept of enhanced view in the year 1950. In the year 1962, the author designed a mechanical prototype, named 'Sensorama,' which utilized multiple senses (sight, sound, smell, and touch) in his five short films. Ivan Sutherland, an American Computer Scientist, in the year 1968, created the first head-mounted display with the help of his student, Bob Sproull. Ivan is also known as the 'Father of Computer Graphics.' The device developed by him was named 'Sword of Damocles.' During the years 1970–1990, VR was widely used in medical simulation, fleet simulation, automobile design, and military training purposes. The term Virtual Reality was coined by Jaron Lanier, an American philosopher, and scientist, in the year 1987. David EM was the first artist to produce a navigable virtual world at NASA's Jet Propulsion Laboratory (JPL) during 1977–1984. Eric Howlett designed the Large Expanse Extra Perspective (LEEP) optical system in 1979. His system created a stereoscopic image with a wide field of view. LEEP is widely applied in photography and VR systems. In the year 1991, Cave Automatic Virtual Environment (CAVE), the first cubic immersive room, was developed by Carolina Cruz-Neira, Daniel J. Sandin, and Thomas A. DeFanti. Further, to provide projection mapping, Michael Naimark, an artist, inventor, and scholar, was appointed as Google's first-ever 'resident artist.' The development of VR technologies over the years is shown in Fig. 7.5.

7.3.2   Categorization of VR

The various form of VR technologies are categorized based on the level of immersion and their applications. As depicted in Fig. 7.4(b), the different VR technologies are as follows:

(a) *Non-immersive simulation*: This is the least immersive implementation of VR. The user enters into the virtual world through a portal, using high-resolution monitors. Conventional desktops or workstations power these types of monitors. In this type of simulation, only a subset of the users' senses is stimulated. The users use keyboards, mouse, and trackballs to interact with the virtual environment. This is the cheapest form of VR technology developed. Virtual Workbench (VW), a non-immersive virtual environment, was developed to understand and interact with complex objects for scientific visualization [110]. Its developers analyzed various use cases such as real-time jet simulator, molecular docking, flow past of tuna, and interaction with molecular dynamics simulations.

(b) *Semi-immersive simulation*: This type of simulation provides more immersive experience compared to non-immersive simulation. The user gets partially immersed into the environment. The semi-immersive simulation system consists of a large-screen monitor, projector, and multiple projection systems. Usually, the resolution of the projection systems ranges from 1000 − 3000 lines. However, to achieve better performance, multiple projection systems are used. A high-performance graphical computing system powers these systems. Semi-immersive simulation has a certain resemblance to flight simulation technologies. Based on a stereoscopic two-sided wall display, a real-time reverse radiosity method [111] was proposed, where its proponents used two-sided walls and L-shaped workbench to create illumination effects, which appear on the surface.

(c) *Fully-immersive simulation*: This type of VR technology provides the best immersive experience to its users. In most of the cases, HMD and motion detecting devices are used to simulate the user's senses. The HMD devices use small monitors in front of each eye to provide stereo, bi-ocular, and monocular images. This type of simulation provides realistic experiences to the user by providing a wide field of view. The users receive a 360° view of the field on moving their head in any direction. In order to achieve a more realistic level of development, the size and weight of HMDs, and system lag may be reduced.

7.3.3   Applications of VR

The various applications of VR technologies [109] are discussed as follows:

(a) Occupational Safety & Health: VR technologies assist in improving the safety of workers at the workplace. Further, VR technologies are capable of testing the product and its viability, and minimize errors before operation of the products. For example, prior visualization of the manufacturing process enables engineers to design and understand the modeling of the system. For prototyping purposes, VR technologies are used during the design phase, which significantly speeds-up the developmental phase. During the training phase, workers get an idea of the real-world consequences of failure.

(b) Medical: In the field of medical science, VR technologies are used for diagnosis and surgeries. Robotic surgery reduces the time and complexities of surgery, helps a human surgeon in performing surgeries, and is often used for training purposes. Additionally, VR is applicable for the treatment of various health issues such as post-traumatic disorder, autistic patients, depression, and phobias.

(c) Training & Education: For training, teaching, and learning, VR technologies are used to provide a 3D view of the object. For example, the 3D view of the solar system and the objects associated with it helps students to impart knowledge and learning. Therefore, VR technologies have resulted in the use of interactive technologies for education instead of traditional pen and paper-based methods.

(d) Entertainment: VR is generally used in various gaming applications such as video games, music videos, fine arts, and 3D movies. The VR HMDs create an immersive environment for the users. The VR glasses are provided with stereoscopic lenses, which enable the users to view 3D objects at different angles. Further, VR is also used in travel and tourism organizations to provide clients with a virtual flavor of possible destinations, cultures, and traditions, before the people actually travel to those places.

(e) Engineering drawing: VR technologies are used in graphics and Computer-Aided Design (CAD) drawings. Clients get a visual idea of the product/-concept to be developed and the model to be implemented in a real-time environment. Further, 2D or 3D view also supports large-scale displays and informations. For example, Autodesk LIVE, an Autodesk technology, helps professionals to transform their models and share designs with minimum delays.

7.4 Big Data and Advanced Analytics

Big data is identified by the huge volume, diverse types, and different sizes of data, which ranges from terabytes to zettabytes. With the integration of advanced technologies such as AI, ML, and IoT to traditional databases, the complexity of data has increased vastly and resulted in the evolution of new forms and sources of data [120]. The processing of big data is important and quite complex because the collected data is mostly unstructured. The various challenges of big data comprises of capturing data, data storage and analysis, data transfer, visualization, data upgradation, information privacy, and data source identification [116]. John R. Mashey, a US computer scientist, popularized the term *big data*, in the 1990s. Further, Kalpan and Haenlein defined big data as *data sets characterized by a huge amount (volume) of frequently updated data (velocity) in various formats, such as numeric, textual, or images/videos (variety)*. The main source of big data is sensors, networks, any type of transactional applications, web, and social media. The categorization of big data based on types, characteristics, and sources are shown in Fig. 7.6.

"Big data will represent the data for which acquisition speed, data volume, or data characterization restricts the capacity of using conventional associated methods to manage successful analysis or the data which can be successfully operated with important horizontal zoom technologies,"– NIST (National Institute of Standards and Technology).

Big data is categorized into three types, which are as follows [118, 119, 117]:

(a) *Structured data*: Structured data is mostly present in an organized format, is easy to analyze, and is explored without any effort, using basic algorithms. Structured data are categorized as quantitative data and are recognized using machine language. These types of data are stored in relational database management system (RDBMS) and are managed by Structured Query Language (SQL). Typically, these data types are generated by humans or machines. Moreover, 20% of the data available throughout the world is structured in nature. Traditionally, most of the business organizations took business decisions based on structured data. Example of structured data includes names, dates, addresses, stock information, and credit card numbers.

(b) *Semi-structured data*: Semi-structured data is more flexible compared to structured data, but less so when compared to unstructured data. Semi-structured data are not generally available in the relational databases. However, these data have certain organizational characteristics that assist in the analysis of data. The scaling of data in the semi-structured format is easier compared to the data in the structured format. Grouping or hierarchy of these types of data helps to assign internal tags and markers. The various formats of semi-structured data are – Markup language, Open standard JSON, and NoSQL. Software platforms such as Apache Hadoop are required to perform the executions on semi-structured data.

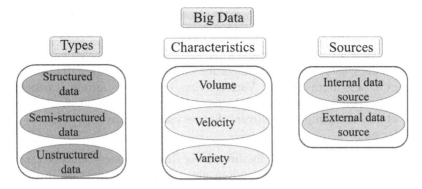

FIGURE 7.6: Big data.

(c) *Unstructured data*: Unstructured data comprises data, which cannot be easily explored applying basic algorithms. These data are categorized as qualitative data. Human-generated unstructured data may be in the form of emails, video files, audio files, social media posts, and images. In contrast, machine-generated unstructured data includes satellite images, scientific data, digital surveillance data, and sensor data. These forms of data are extracted through an object-oriented database such as Apache Hadoop. Around 80% of the data available online are unstructured. In the case of huge datasets, unstructured data provides better insights. The unstructured data can be used to analyze digital communications, track high-volume of data, and gain intelligence through analyzing human behavior by the use of advanced tools such as ML.

7.4.1   Characteristics of big data

The data generated from intelligent systems in the IIoT environment are colossal in amount, vary in type and nature, and data formats. Therefore, big data has three specific characteristics – volume, velocity, and variety. However, many researchers also believe that big data may be characterized by variability, veracity, value, and visualization. In order to extract meaningful information from the data, it is necessary to analyze the data. The analysis of big data is complicated because of the diverse nature of the data being handled [120, 121].

- Volume: After the metamorphosis of the traditional industries to Industry 4.0 and IIoT compliant, heterogeneous types of sensor nodes attached to the machines and various devices are quite common. These devices generate a huge amount of data. The size of data may vary with time and range from a few kilobytes to zettabytes. For example, the large synoptic survey telescopes (LSST) generates almost 32TB of data every night. Similarly, in another context, about 15 million customers click every day to provide feedback on the Amazon webpage.

- Variety: As the amount and type of data generation characteristics are random; therefore, the organization of this huge volume and diverse form of data is quite difficult. Variety refers to the data format – structured, semi-structured, or unstructured – in which the input data is generated. Furthermore, intelligent devices can generate data in any of these formats.

- Veracity: The veracity feature of big data indicates the correctness of the data. Based on the random nature of data generation in terms of volume and type, accuracy analysis of data is complicated. Industry 4.0 compliant industries comprise of smart machines, which are capable of self-diagnosis of faults and maintenance. Therefore, to provide real-time feedback to the machines, advanced learning algorithms analyze these generated data. The veracity of big data makes the analysis of data complex.

- Velocity: This characteristic signifies the speed at which the data is generated. With the advancement in technologies, the data processing speeds of processors have increased. However, the overall performance of systems have reduced significantly due to huge data volumes being handled at each time instant, especially as big data requires the handling of data in real-time. The frequency of generation and handling the data may vary. For example, 140 million tweets are produced on average per day, as per a survey in 2011.

- Visualization: The big data generated can be visualized in the form of pictures and graphics. Based on this visualization, decision-makers can formulate new strategies and have a complete picture of the industrial landscape of their choice. Any new pattern in the data format can be identified using this characteristic of big data. Therefore, the appropriate utilization of the data generated indicates the pattern, trend, and correlation in the data.

- Value: The value of big data indicates the process of analysis and extraction of meaningful information from the generated data. The process of computing the value of data is simple. Further, the value extracted assists in the evaluation of the investigation process and helps the concerned personnel to make a decision.

- Variability: The variability characteristic of data signifies the variation in the meaning of data. However, data variability is different from data variety. The randomness feature in the generation of data indicates the variability characteristics. Therefore, the process of handling data becomes difficult due to inconsistency in the data. For example, language processing, hashtags, geo-spatial data, and multimedia sensor data.

7.4.2 Big data sources

Big Data generated from the devices across various industries can be categorized as *internal* and *external* data sources. Internal data sources represent the data generated from the devices of an organization, which are also owned by them. Enterprise data, IoT data, biomedical data, and computational biology are the different forms of internal data sources. On the other hand, external data sources represent publicly available data sources that are not owned by an organization [124]. The data generated from social media, publicly available datasets used for ML, and weather forecasts are some examples of external data sources.

(a) Internal data source: The frequency of data generated from the sensors attached to different interconnected IoT devices serving various industrial applications such as industrial (manufacturing and non-manufacturing industries), agriculture, and healthcare data, vary with time. Further, based on the type of sensor used, the quality of the data produced changes. Any financial transaction, production and inventory data, and online trading and analysis generate data either in the structured or unstructured format. The organization, processing, and analysis of these collected data are based on the quality of information produced from business transactions. Similarly, experiments on gene sequencing, computational biology, and physiological data of patients [122] are also producers of a huge volume of data.

(b) External data source: Typically, a huge amount of data is generated from different social media such as Twitter, Facebook, and LinkedIn, and weather forecasts. This type of data produced may be quantitative or qualitative. The qualitative aspect of data includes the accuracy of the analytics, and are written in natural language. On the other hand, the quantitative aspects comprise the number of observations categorized based on geographical or temporal characteristics. Publicly available datasets are significantly diverse for performing advanced analytics. In addition to this, the traditional and modern databases also act as external sources of big data [125].

7.4.3 Big data acquisition and storage

The acquisition of data involves the procedure for collection, transmission, and pre-processing of data. Based on their characteristics, the acquisition of big data is handled accordingly in the industries. The acquired data is processed, analyzed, and stored in data warehouses. Big data acquisition is made through message streaming protocols, also known as a publish-subscribe model. The basic assumption of big data acquisition is to collect the data from multiple data sources, store them, and analyze them through the application of various protocols. In the preliminary step of data acquisition, data is accumulated from various sources and temporarily stored in a database or file. Further, the collected data is categorized as either intra- or inter-data center transmitted data. Data that are collected and interconnect information internally within a data center are referred to as intra-data center data. On the other hand, the data connecting between the data centers are known as inter-data center data. These data are processed and analyzed by applying big data analytics. The pre-processing of data comprises integration, checking the correctness of data, and removal of redundant information from the data. Data integration means a combination of data from different sources to provide a clear view to its users. After integration, the quality of data is checked through identifying and removing incorrect or insufficient data. Finally, to avoid unnecessary transmissions, the redundant modules in the data are removed through filtering and compression [123]. The three basic components of big data acquisition are as follows:

- Different protocols, which are used to accumulate data from various sources and forms, and convert them into a meaningful format. For example, the Advanced Message Queuing Protocol (AMQP).

- The frameworks which help in the collection of data from various sources and apply the protocols.

- Storage of the data accumulated by the frameworks.

With the increase in popularity of social media, the amount, nature, and type of data produced, ranging from sensors to online transactions, have increased significantly. Big data widely influence almost all business sectors. The methods applied to collect, process, analyze, extract meaningful information, and store big data depends on the characteristics of the data. The storage of the colossal amount of data produced from different sources, which serve a diverse range of applications, is a difficult task. A distributed file system is utilized to store these vast data such that consistency, fault-tolerance ability, and accessibility of data is maintained. Different projects such as Hadoop, Google File system (GFS), and Flumes are used to store the data. Hadoop is an open-source project, which is reliable and scalable. Hadoop Distributed File System (HDFS) is a crucial file system, which reliably stores terabytes and petabytes of data. Typically, HDFS is suitable for processing unstructured data. A master/slave architecture[¶] is followed by HDFS to communicate between devices. The HDFS architecture complies with a hierarchical system. In order to access these massive data efficiently and reliably, the GFS is employed. Flumes is another form of service used to gather and transfer big data efficiently in a simple manner. Further, non-traditional relational databases

---

[¶]In the master/slave architecture, one or more devices can communicate with other devices/processes in a unidirectional manner. In a system, the master node selects other devices. The devices which act according to the master are known as the slave devices.

(NoSQL) are used for storage and retrieval of big data. As compared to the traditional data models, NoSQL allows flexible design of schema during the upgrade of the database. Moreover, NoSQL can manage unstructured data and brings scalability to the system.

7.4.4 Necessity of data analytics

With the introduction of Industry 4.0 and IIoT, heterogeneous types of sensor nodes are being attached to the machines/devices. These smart sensor-enabled nodes sense and transmit data to a server using wireless technologies. After that, the categorization of the data, storage, analysis, and elimination of redundancies from the data is performed. Additionally, as these data are collected from multiple sources and in diverse forms, data analysis is necessary to process this massive amount of data and classify them accordingly. Data analysis is applicable in various fields such as business, medical/healthcare, transportation, social science, and manufacturing industries.

In the various business sectors, data analysis helps in decision making. Further, data analysis helps to forecast the future trend of customers' demands, analyze the data in a meaningful way, and increase productivity in businesses. The suggestion of appropriate products to customers using data analytics helps businesses to have wider reachability and sell more products, which in turn translates to more revenue for these businesses. Similarly, data analysis can predict the future occurrence of faults, extract meaningful information from the data, help in improving the safety of workers, and enhance the overall efficiency in the manufacturing and process industries. Further, real-time analysis of the medical data of a patient helps to predict their health conditions from stored electronic records. Finally, in the field of transportation, analysis of the data collected from sensors may result in the reduction of accidents, injuries, and fatalities. Further, data analysis may help in the selection of the shortest route by the driver, which leads to fuel efficiency and saves time.

Traditionally, DWs act as the central repository of data collection, information retrieval, and analysis. DWs are central repositories, which store and collect data from one or multiple sources within a specified period. However, big data analytics is different from DWs. As a massive amount of data flows in real-time, a parallelly performing and scalable framework is necessary for big data analysis. Cloud-based analytics is a strong contender for handling such data. According to NIST standards, the essential features of cloud-based analytics are as follows:

- On-demand self-service: The services are remotely provided to the customers/end-users on a payment basis, as per their requirement. The customers remain unaware of the back-end process of how the services are provided to them and require minimum interaction with the service provider.

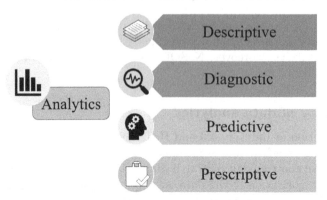

FIGURE 7.7: Types of analytics.

- Network access: The services are available over the network and are easily accessed by consumers through standard mechanisms.

- Grouping: Based on the demand of consumers/end-users, these services are grouped. Further, the resources are dynamically provided to the end-users, on request.

- Flexibility: The services requested by consumers are almost equivalently provided to them in appropriate quantity and time. However, these services may appear to be immense to consumers.

- Measured services: The type of service and amount of resources used at any level can be optimized and controlled with proper clarity by both the service provider and consumers.

7.4.5 Types of analytics

Analytics is the combination of various processes such as information extraction, transformation, and inferring from the data collected to make a decision. Advanced analytics help to improve the process of decision making. Big data can be applied to reveal hidden patterns and information/s from the accumulated data. Fig. 7.7 shows the different types of analytics. The various types of analytics are as follows:

(a) Descriptive analytics: The analysis of *what happened* in the event is performed by applying descriptive analytics. This analysis utilizes historical data to analyze and provide insights into events. For example, the sales of a company can be predicted from the preference of customers and the sales cycle.

(b) Diagnostic analytics: Diagnostic analytics infer from historical data and find dependencies. The detailed analysis of *what is likely to happen*, and the pattern can be identified from the data. For example, what are the probable reasons for any particular disease, a patient is suffering can be predicted from the past data. Further, the future sales and profit of a company can be estimated from the previous sales data.

(c) Predictive analytics: Predictive analytics help to predict the probability of *what will happen* in the future. The results of the descriptive and diagnostic analytics are used to predict future trends, the correlation between the events, and detect similar clusters. For example, analysis of sales source, lead time, number and type of communications, and social media documents, may help to forecast and improve the sales of a company.

(d) Prescriptive analytics: The measures undertaken to eradicate issues predicted and finding the correlation among them to prove/nullify a hypothesis is enabled using prescriptive analytics. Both the historical data, as well as other information, are combined for enabling prescriptive analysis. For example, identification and clustering of patients with similar LDL, cholesterol, and high-pressure levels.

7.5   Smart factories

The transformation from the traditional industrial system to a connected, automated, self-corrective, and flexible system is the major driving factor towards the development of smart factories. The primary motivation behind the evolution of the concept of smart factories is – fulfill the demand of customers, improve the product quality, and enhance the safety of personnel. Automation has existed in the control systems of industries for the past few years. However, the rapid rise in technologies such as AI, robotics, big data analytics, end-to-end connectivity, and digitization, has become an essential component of smart factories. Further, the complexity in the supply chain has increased due to the fragmentation of various processes among multiple suppliers across the globe. Automation in the various units of factories due to the integration of operational and information technologies act as an important factor in the upgradation of traditional factories to a smart factory. This transformation in the industrial sector will result in the transformation of linear supply chains into an interconnected digitized system. The amalgamation of the advanced technologies with the existing system results in the development of a smart factory across the industry and the entire supply chain. Fig. 7.8 shows the schematic diagram of a smart factory. The smart sensors, machines, and devices sense and transmit data to the edge/cloud. After that, the complex analysis is performed at the cloud, as shown in Fig. 7.8. The transformation in

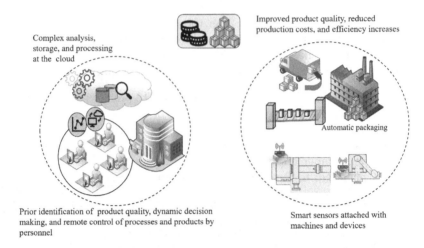

Complex analysis, storage, and processing at the cloud

Improved product quality, reduced production costs, and efficiency increases

Automatic packaging

Prior identification of product quality, dynamic decision making, and remote control of processes and products by personnel

Smart sensors attached with machines and devices

FIGURE 7.8: Schematic diagram of a smart factory.

factories results in an efficient system, less production downtime, improvement in product quality, and enables factories to self-adapt with the environment.

> The smart factory is a flexible system that can self-optimize performance across a broader network, self-adapt to and learn from new conditions in real or near-real time, and autonomously run entire production processes [126].

The integration of new technologies and upgradation of the entire system to a smart factory and digitized supply chain may seem complicated. However, keeping with the present-day trends, it is a crucial requirement to compete in the market and improve production in addition to scaling up the whole system. In this new supply chain, various entities can communicate and establish connectivity among the unconnected links in the digital network. Moreover, smart factories are capable of operating in almost real-time and develop the requirements of the customers.

The various advantages of a smart factory are as follows:

(a) Reduction in overall production costs: Traditional processes require higher levels of inventory, human resources, and have less variation in the operations. With the advancement in technologies, both the quality and quantity of the produced goods improve. Additionally, operational and maintenance costs are reduced significantly in the long run.

(b) Increase in efficiency: Smart factories are self-adaptive and self-corrective to continuous analysis, promote optimization of processes, reduce downtimes, and improve the capacity of plants.

(c) Improvement in product quality: The integration of advanced technologies results in a reduction in product defects. Through early prediction measures and corrective actions, the product quality is improved and maintained.

(d) Upgradation of safety in the factory: Safety of the workers is one of the key aspects of Industry 4.0. Further, as the advanced technologies require minimum or no human intervention in the production process, they result in lowering the percentage of errors and defects in the products.

(e) Enhanced prediction of errors: The application of predictive and prescriptive technologies on the real-time data identifies hidden patterns in the data and improves the efficiency within the supply chains.

7.5.1 Characteristics of smart factory

The transformation of traditional industries to smart factories requires the adoption of certain changes in the existing processes. The ability to predict from real-time data and dynamically adapt these changes result in the improvement of overall efficiency in a smart factory. Various characteristics of a smart factory are as follows:

(a) *Connected environment*: One of the important features of a smart factory is connectivity among the various devices or things. In a smart factory, the raw materials, resources, and information are interconnected and designed to operate in real-time. On the other hand, smart sensors form the basic skeleton of the industrial processes. The data sensed by these sensor nodes are integrated with the different units of a factory. For example, operation and business systems interconnect and communicate with other units. This results in communication among the various entities of the supply chain and provides real-time status of each sector of the chain, which improves the efficiency of the entire supply chain.

(b) *Optimal utilization of resources*: The fourth industrial revolution enabled the machines to communicate among themselves. The primary objective of M2M communication is to reduce errors, enhance safety, improve the plant life-cycle, and minimize wastes. A smart factory optimally utilizes the resources with minimum or no human intervention, increasing uptime of machines, improving the quality of products, and optimization of the energy consumption.

(c) *Transparency of process status*: The real-time data collected by sensors are processed, and meaningful information is extracted from them. Further, this assists in the prediction of the number of products to be produced from the factory and satisfies customers' demand in the future.

(d) *Dynamic decision making*: Smart factories have an interconnected environment, such that any form of abnormality in the processes is detected and resolved in real-time. There are certain essential aspects in smart factories, which require dynamic decisions, such as the automated restocking and replenishment of inventory, identification of the quality of the product before delivery from suppliers, and safety issues. The manufacturers and other personnel are capable of digitizing an operation and integrating the predictive qualities with them.

(e) *Flexibility*: This is one of the salient features of smart factories. Various process parameters, such as the configuration of machines, the flow of materials, and temperature control, can be optimally set in a smart factory. Therefore, flexibility in the machines results in an increase in production, reduced downtime, minimizes errors, and helps in scheduling the processes in the plant.

7.5.2   Technologies used in smart factories

With the incorporation of smart technologies, the factory units are capable of interacting with one another and predict the anticipated changes. The various technologies used in smart factories are as follows:

(a) Product life-cycle management (PLM): PLM assists in the management of the entire life-cycle of the product, from design to the delivery of the product. Through the entire process of PLM, data is collected, processed, and analyzed at every step to help improve the growth of industries. With the help of PLM technologies, integration among the products, people, data, and processes becomes easier.

(b) Execution systems in manufacturing: The manufacturing execution systems comprise of smart machines, smart devices, and smart manufacturing processes. Sensor nodes are attached to these devices, which sense and transmit data to the edge devices or cloud for further processing. These advanced devices help to improve productivity, reduce wastes, and minimize the overall production costs.

- Smart machines: Smart machines communicate among themselves and other devices within their proximity. These machines are capable of adapting to the environment, correct errors, analyze them, and takes corrective actions. Further, machine-to-machine communication also enables humans to interact with the machines.

- Smart devices: Smart machines connect with smart devices, which assist in real-time data collection, analysis, and communication. The smart devices result in improved utilization of assets, minimize energy consumption, reduce maintenance costs, and improve efficiency.

These devices communicate with other devices through various wireless protocols such as Bluetooth, NFC, and Wi-Fi.

- Smart manufacturing processes: With the advancement in technologies, the productivity of industries and the efficiency of the manufacturing processes have improved significantly. Smart manufacturing processes provide real-time information, make dynamic decisions, reduce latency, and enable efficient communication among the different processes [127].

(c) Industrial Automation: Industrial automation helps in reducing human labor, collect data from processes, and improve the safety of workers. Automated manufacturing improves the quality of the products and reduces errors in the system. The transformation of traditional machines to smart machines and smart manufacturing processes results in the replacement of humans by machines in hazardous locations of the factories.

- Data analysis: Automation in the industrial processes is achieved utilizing the real-time data collected from sensors embedded into the devices. These data are processed and converted into a meaningful form, after which decisions are generated from the data, which helps in the predictive forecast of various process parameters. In the Industry 4.0 environment, AI and various complex learning methods are used for the analysis of data.

- Information technology: Information technology plays an important role in transforming traditional industries into smart factories. Smart software applications, control, monitoring, and management are facilitated by smart sensors, smart meters, and intelligent devices. The data sensed and collected by these sensors, and other devices are processed, analyzed, and stored in production management logic using intelligent production management software.

7.6 Lean manufacturing system

The concept of lean manufacturing system was introduced by Toyota to minimize wastes from its processes and efficiently utilize the resources in their plants [129]. Through the elimination of wastes from industrial processes, the lean approach helps to improve efficiency, optimize cycle time, enhance productivity, reduce response times, reduce material costs, and facilitates high throughput. As wastes are reduced, the quality of the product improves, and the overall production time and product costs reduce. The primary aim of the lean approach is to focus on the needs of the customers. Fig. 7.9 shows the main aim behind lean production, different types of wastes in the factory, and

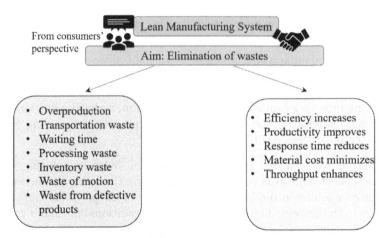

FIGURE 7.9: Lean manufacturing system.

outcomes of lean production. However, other existing approaches related to the various operation and controls in industrial sectors are considered from the perspective of tasks and production. Further, the lean approach is applicable across various industrial sectors, such as the manufacturing sector, inventory management, and healthcare industries.

The lean manufacturing system is based on two approaches – JIDOKA and Just-In-Time (JIT) approach. JIDOKA helps to eliminate the main reason for the production of wastes. The JIDOKA concept describes that in case of any abnormalities detected by the machine, it stops operation to prevent the production of defective products. Further, smart machines are capable of operating autonomously with minimum or no human intervention. In other words, a combination of human intelligence and automation lead to the initiation of JIDOKA. JIDOKA and JIT are combined together to improve product quality, enhance the efficiency of the plant, and reduce wastes. JIT helps in continuously improving the performance and quality of products while enabling flexibility to the system. This approach assists producers in forecasting demand. Therefore, to satisfy customers' demand, an adequate amount of inventory is maintained in the stores by the producers [128]. The essential elements of JIT are – a continual improvement, waste elimination, maintenance of hygiene and cleanliness at the workplace, improving flexibility by producing in small batches, and taking corrective measures during faulty operations.

The elimination of wastes is one of the main objectives of the lean manufacturing system. There are seven types of wastes produced, which are as follows:

(a) Overproduction: Producing a product before it is demanded by customers may result in the degradation of product quality and productivity of the unit. Overproduction results in an increase in lead time, increase storage costs and hinder the detection of errors.

(b) Transportation waste: The transportation of products from the production unit to the distribution centers adds transportation cost to the product price. Additionally, excessive movement and mishandling may result in deterioration of product quality. Proper mapping of the products to the delivery locations may reduce the transportation cost.

(c) Waiting time: As the waiting time of a product increases during processing, material flow degrades, and lead time increases. The period of inactivity of people and the flow of information or products result in waiting time. However, linking the processes appropriately, such that one process outcome is fed directly to the next process, may result in a reduction of waiting time.

(d) Processing waste: The use of inappropriate tools for processing may lead to an increase in the cost of goods produced. However, using small and flexible equipment for production may reduce the waste of inappropriate processing.

(e) Inventory waste: The presence of excess inventory on the factory floor reduces space, acts as a barrier to effective communication, increases lead time, and introduces delays in the identification of faults. The costs associated with the wastage of inventory is reduced by proper execution of inventory flow among different work centers.

(f) Waste of motion: The non-value added movement of working personnel such as walking, bending, stretching, and lifting on the factory floor exposes them to health and safety issues. Therefore, jobs with excessive motion may be redesigned to avoid motion wastes.

(g) Waste from defective products: The total cost of defective products produced by an organization form a greater percentage of the manufacturing costs. However, through the active participation of the workers and continuous improvement programs, the wastes produced from defective products are reduced.

7.6.1 Value streams in lean production system

A *value stream* refers to the set of activities taken up by an organization from order to the delivery of products. The primary objective of a value stream is to concentrate on the areas which add value to the product. In order to predict the future state of certain events, the appropriate analysis of the present state is necessary. The structured steps of a lean management approach are used to map the future state, improve the flow of the product and information, and reduce wastes. This is known as *value stream mapping*. Therefore, the efficiency of the entire system improves [130]. The steps to be followed to create an effective value stream are as follows:

(a) Selection of the appropriate value stream is necessary. Based on the objectives for a specific product, value streams are chosen.

(b) The identification of the proper personnel to be included in the team, who are capable of making necessary changes during the development of the product.

(c) Another important aspect of value stream mapping is to visit the workplace, list the ideas from a customers' perspective, and work upon these ideas.

(d) Scheduling date and time, and coordinating with the selected team is also an essential part of value stream mapping. Based on the mapping of the activities with the team members, the wastes associated with money and time may vary.

(e) Lastly, summarizing the steps to the team members may result in saving the time wasted for understanding the activities.

7.6.2   Necessity of lean production system

The lean manufacturing system has attained popularity across various industrial sectors. With the application of the lean approach in plants, the overall productivity and quality of products have improved. Additionally, the amount of wastes and the cost of the product is reduced significantly through the use of lean approach. The application of advanced information and communication systems developed during Industry 4.0 and IIoT plays a vital role in the development of smart factories. Additionally, the lifetime of machines improves, downtime of machines reduces, and overall production increases [131]. Therefore, to integrate a lean approach with Industry 4.0, the concepts of JIDOKA and JIT play an essential role.

The appropriate usage of inventory, proper synchronization among the manufacturers and suppliers, and reallocation of tasks among workers act as essential factors in the integration of a lean approach with the industries. Customers form the core of a supply chain. Therefore, analysis of customers' demand and their satisfaction play an essential part in the industries. There are specific process parameters, which are responsible for value stream mapping in the factory floor, such as the amount of inventory present, machine-to-machine communication, machine-to-human communication, and flow of raw materials. Further, the predictive and prescriptive analysis of the real-time data available helps in taking corrective actions. A lean approach is essential for establishing efficient machine-to-worker communication, improving the overall control system, enhancing the coordination among different entities of a supply chain for proper flow of products, and improving the working conditions of workers. With the adoption of a lean approach, the profit and quantity of products produced increases for the same level of investment.

7.6.3 Implementation of lean manufacturing system

Lean manufacturing system can be represented as a fixed state, continuous improvement, set of working methods, and philosophy for application. The implementation of lean approach is not subjected to lean tools. To implement a lean approach, the process operators need to select the appropriate technical team for providing a solution to problems. After that, proper mapping of the value stream is vital to reduce wastes. Communication among the various levels is essential for the implementation of the lean approach. In case of any crisis, a long-term focus on the step-by-step upgradation may help in the process of lean transformation. Further, real-time data is collected, processed, and analyzed for systematic review [133].

The four primary areas for implementation of lean approach are – business requirements, operation improvement, people management, and performance governance.

(a) Business requirements: In order to successfully implement lean approach in businesses, organizations set the right objectives at the outset, develop strategies, and decide the contributions of team members developed.

(b) Operation improvement: With the help of appropriate knowledge of tools and processes, an operation can be improved.

(c) Performance governance: The performance of an organization is guided with the help of certain factors termed as key performance indicators (KPIs). Through a clear understanding of the processes, the quality of products, safety of personnel, and overall efficiency can be improved.

(d) People management: The training of workers helps to provide them with clear knowledge of advanced technologies. Further, they develop capabilities to perform their work properly.

Summary

This chapter introduces some of the key technologies necessary for Industry 4.0 and IIoT, such as AR, VR, Big data and analytics, smart factories, and lean manufacturing system.

(a) The transformations required for the development of Industry 4.0 and IIoT are – upgradation of machines, smart sensor nodes, incorporation of new business strategies, improvement in technical skills of workers, and incorporation of changes in factories.

(b) AR combines real-world and computer-generated effects to provide a magnified view of the physical world.

(c) Based on the technologies used and application-based use cases, AR is categorized as Marker-based AR, Markerless AR, Projection-based AR, and Superimposition-based AR. Moreover, AR is widely applicable in diverse fields such as military, healthcare, supply chain, entertainment, and education.

(d) VR generates a realistic 3D artificial environment, which provides users with an immersive experience.

(e) Based on the level of immersion, VR is categorized as non-immersive simulation, semi-immersive simulation, and fully-immersive simulation. VR technologies are widely used in various applications such as OSH, medical, training, entertainment, and engineering drawing.

(f) The analysis of the huge amount of data generated in the IIoT environment is difficult because of its volume, variety, velocity, veracity, value, variability, and visualization.

(g) In order to process, store, and analyze the big data collected from industries, different protocols and frameworks are applied.

(h) Smart factory is the result of the transformation from the traditional industrial system to a connected, automated, and flexible system.

(i) JIDOKA and JIT approaches are combined to facilitate the lean manufacturing approach. The primary focus of the lean approach is waste reduction and upgradation of product quality supplied to customers.

(j) In order to apply lean approach, the selection of appropriate value streams, proper use of inventory, and allocation of tasks among the workers are necessary.

Exercises

1. What are the transformations required in the industrial processes to become Industry 4.0 compliant?

2. What is AR? What are the hardware components involved in AR technologies?

3. What are the key features of AR?

4. In which year was the term AR coined? What was developed in the US Naval Research laboratory?

5. What are the various AR-based technologies? Discuss one of them.

6. Write two differences between marker-based and markerless AR technologies.

7. In which military applications AR is applied?

8. How AR technologies are applied in the field of healthcare?

9. What is VR? What are the key features of VR?

10. What do a semi-immersive solution comprise of?

11. How VR technologies are applied for training and education?

12. What is big data? How is big data categorized? Give some examples of each type.

13. What are the various sources of data? Explain.

14. What are the basic components for the acquisition of big data?

15. What are the necessary reasons for the application of data analytics?

16. What are the features of data analytics?

17. What are smart factories?

18. What are the different types of analytics methods?

19. What are the advantages of smart factories?

20. What are the characteristics of a smart factory?

21. Discuss different technologies applied in smart factories.

22. What is the primary objective of a lean manufacturing system?

23. Discuss the different types of industrial wastes.

24. What is a value stream? How to create a value stream?

25. What are the four application areas of the lean manufacturing approach?

26. What are the necessary human skills required to be developed in various industrial sectors to become Industry 4.0 compliant?

27. In order to provide an enhanced view of the real world, what type of sensory modalities are produced by AR technologies?

28. Which AR technology is known as position-based AR? Why is it called so?

29. In which VR technology motion-detecting devices and HMDs are used?

30. What are intra- and inter-data centers?

31. How is a lean manufacturing system implemented in industries?

32. What are virtual fixtures? When and where was it developed?

33. What is CAVE? Discuss in brief.

34. Write the differences between structured and unstructured data.

35. What is a value stream mapping? Discuss a case study on value stream mapping in industries.

Enabling Technologies of IIoT

8

Sensors

Learning Outcomes

- This chapter provides an introduction to the various sensor types and their underlying working principles.

- New readers will be able to have a grasp of the terminologies, technologies, and requirements while selecting sensors for their applications.

- Seasoned learners will be able to appreciate the overview of the various sensor types available, which will help them narrow-down on a sensor or sensor type for their solutions.

8.1 Introduction to Sensors

The present-day consumer, as well as industry-grade devices and machinery, are equipped with at least one sensor [134]. The presence of multiple sensors on a single device is quite normal nowadays. The big question is, what are these sensors, and why are they being used with increasing intensity and quantities as the devices become more smart and autonomous? The answer to this lies in the comparison of any of these devices with a human body. Humans have evolved to have the five basic senses of touch, sight, hearing, smell, and taste. These senses help a human being to feel physical sensations (pressure differences), see around them (light intensity and frequency differences), hear their surroundings (sound/acoustics), differentiate between materials based on their aroma/smell (chemical), and mostly distinguish between edible items by tasting them (chemical and texture). Now, comparing these features of a typical human with any modern-day device such as a smartphone, we can easily see the similarities and form an idea about the importance of sensing in various operations, especially industrial ones. A cheap smartphone can be found with sound sensors (microphone), light sensors (controls the display brightness of the screen automatically), color sensors (camera), and pressure sensors (barometers and touch screen). Advanced smartphones have magnetic sensors, acceleration sensors, temperature sensors, impact sensors, and other features depending on the price range.

Traditional industrial processes are rapidly transforming towards Industry 4.0 standards by replacing processes consisting mostly of mundane and repetitive tasks into highly automated processes [135]. The core of these automation tasks consists smart processors and algorithms, which, however, are facilitated by sensors [136]. Starting from regular industries – agriculture, transportation, logistics, and others – to highly specialized and critical ones – chemical plants, thermal plants, and others – the type, quality, and the number of sensors vary according to the sensing requirements.

Sensors are used to quantify physical phenomenon (changes in pressure, temperature, light intensity, sound pressure, and others) into electrical signals. These generated signals are used for determining and characterizing various forms and stages of physical processes and phenomena into a human-understandable format [137]. Once, a physical phenomenon becomes human-understandable, its characterization and control can be planned. For example, temperature sensors can be used to determine the boiling point of water. However, the same conditions (amount of heat, duration of heating, ambient weather, ambient pressure), which were required to make the water boil, cannot always be recreated. In such scenarios, the quantification of water's boiling point as $100^{o}C$ by a temperature sensor is fixed and universally standard. This quantification of water's boiling point is the same, even on high mountains as well as plains.

Typically, regular industries require a fixed number of sensors, which can work under normal operating conditions, temperatures, and pressures. However, specialized industries such as foundries, chemical plants require sensors, which are capable of quantifying extreme physical conditions to signals. The presence of high temperature, high pressure, and corrosive materials are commonly associated with such industries.

8.2   Characteristics

Sensors are characterized by a few traits, which may be inherent to the material using which they are fabricated, or the design form the sensor is being as, or even due to manufacturing defects. The typical sensor characteristics are described in the subsequent subsections.

8.2.1 Sensor calibration

This feature defines the method chosen to remove sensor errors, which have inherently been incorporated in the sensor during its design. It denotes the deviation of a sensor's output values from the expected values, which are expected from an ideal sensor of the same type, operating under ideal conditions. Fig. 8.1 shows the graph for calibration of sensors. Typically, calibration is used for weeding-out repeatable errors. Through characterizing the nature of these repeatable errors, sensor readings can be corrected by applying the correct compensation to the measured readings against the characterized errors. Calibration is a means of enhancing the accuracy of sensors under real-life operating conditions.

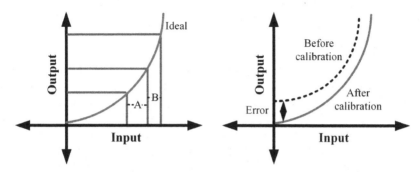

FIGURE 8.1: A typical sensor calibration graph.

8.2.2 Sensor profile

This feature defines the expected trend and behavior of a sensor being used. Sensors can be profiled as either linear or non-linear. Sensors are designed in such a manner that the output of a sensor varies proportionately with the input parameters. The input parameters can be temperature, voltage, current, or any other characterizable parameter. If the output is directly proportional to the input, the sensor is categorized as a linear sensor. Fig. 8.2(a) shows the characteristics of a linear sensor. In contrast, Fig. 8.2(b) shows a non-linear sensor's characteristics. In other words, for a fixed change in the input value, the sensor output always changes by a fixed value, a linear behavior is observed between the input and output. In contrast, if the output varies in a manner that the constant of proportionality between the output and input values is not linear (polynomial, exponential, square, logarithmic, and others), the sensor is categorized as a non-linear sensor.

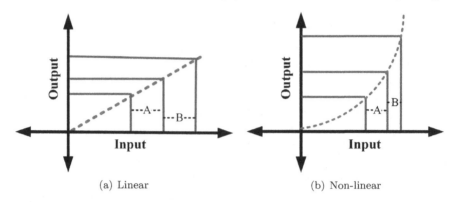

(a) Linear   (b) Non-linear

FIGURE 8.2: Typical sensor characteristic graphs.

8.2.3   Sensor accuracy

The accuracy of a sensor is defined as the degree of correctness or closeness with which a sensor can quantify a physical phenomenon. For example, consider a vat of boiling water (water boils at 100^oC). A temperature sensor T_1 outputs a value of 99.8^oC, whereas T_2 outputs 103.1^oC. We say that T_1 is more accurate than T_2, even though they do not show the exact correct value. If we observe closely, T_1 deviates from the actual temperature of 100^oC by 0.2^oC, whereas T_2 deviates by 3.1^oC. Therefore, T_2 is prone to larger deviations from original values as compared to T_1.

8.2.4   Sensor resolution

The resolution of a sensor signifies the intensity or level of details a sensor can capture from a physical phenomenon while quantifying it. For example, consider an accelerometer, which outputs the value of acceleration (in this example, along the x-axis only) in m/s^2. This sensor a_1, along with another similar sensor a_2 is attached to a car accelerating at 3.2 m/s^2. The first sensor a_1 reports the car's acceleration as 3.225 m/s^2, whereas the second sensor a_2 reports it as 3.22 m/s^2. Here, we say that the resolution of a_1 is more than a_2 as it can provide measurements of the phenomenon with more precision (up to 3 decimal places) as compared to a_2 (up to 2 decimal places).

8.2.5   Sensor rating

The rating of a sensor signifies the maximum and minimum values of the different conditions (both ambient as well as input), which the sensor can safely handle without damaging itself or distorting its readings. A non-contact infrared thermal sensor is rated to operate between -10 to 200^oC. Operating this sensor to measure temperatures greater than the maximally rated value of $200^{\circ}C$ will saturate its readings by overloading its sensing elements. This will result in incorrect readings. Fig. 8.3 shows a typical sensor rating graph. The input and output characteristics can be plotted against various sensor parameters. The plot shows the variation in the sensor's output values for variations in the sensor's input values. It is to be noted that the sensor output is truly active for a fixed range of sensor input values, which is known as the sensor's rating.

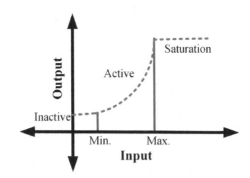

FIGURE 8.3: A typical sensor rating graph.

8.2.6 Operating voltage

This feature signifies the range of operating voltages, within which the sensor can operate without damaging its (associated circuitry). It is to be noted that this feature is typically present in sensors which require an external source of power to operate/function. Generally, commercial sensors available for circuit enthusiasts and hobbyists have an operating voltage range of 3.3 to 5 V. Operating this sensor to supply greater than the maximally rated 5 V increases the chance of damaging the sensor. Additionally, operating this sensor below the minimum value may not be enough to activate the sensor at all. Fig. 8.4 shows a typical sensor output graph, plotted against varying input conditions. The plot shows the minimum and maximum values of inputs, which can be applied to the sensor to generate a proper response from the sensor.

8.2.7 Output

This feature signifies the nature and range of output a sensor will provide. The output of a sensor can be either analog or digital, each of which requires separate circuit configurations to read, even if both these different sensors (analog/digital) measure the same parameter. Additionally, the output of the sensor may also be encoded and require special circuitry or software

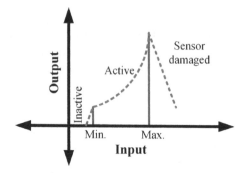

FIGURE 8.4: A typical sensor output graph.

procedures to decode or read. For example, a commercially available temperature sensor can be of two types—analog (LM35), and digital (DHT11). The LM35 sensor positively requires an analog-to-digital converter before its values can be understood by a processor to which it is connected. In contrast, the DHT11 sensor can directly connect to a processor. However, it requires a special software routine to be incorporated at the processor to make the processor understand its readings.

8.3   Sensor Categories

Sensors perform a wide variety of tasks in many domains and industries. Most modern-day industries rely heavily on sensors for keeping their processes under control, automating tasks, and even managing mundane tasks such as stock-keeping and record maintenance. We categorize various classes of sensors depending on their underlying sensing task into six categories in the subsequent sections.

8.3.1   Thermal sensors

This class of sensors is used for quantifying temperature or thermal energy [138], [139]. These sensors may be designed to output values in Celsius (^oC), Fahrenheit (^oF), Kelvin (K), or Joules (J). It is to be noted that heat and temperature are separate entities. Heat is the energy transferred between molecules, especially during various chemical reactions. As heat is a measure of change, a body cannot have its own heat; it can either gain or lose heat. In contrast, the temperature is the average kinetic energy of the molecules in an object, and is measurable. Thermal sensors can be broadly classified into three types;

(a) Resistive Temperature Detectors (RTDs): This sensor measures absolute temperatures with very high precision. It uses highly pure forms of conducting materials such as platinum and copper to develop changes in the material's resistivity due to changes in temperature. However, these types of temperature sensors have reduced sensitivity to temperature changes. RTDs are passive devices as they need an external voltage supply to operate (Fig. 8.5).

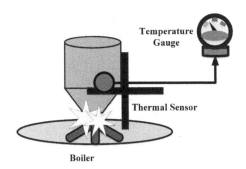

FIGURE 8.5: A typical thermal sensor arrangement used in detecting temperature changes in an industrial boiler.

(b) Thermistor: This type of temperature sensor is a thermally-sensitive resistor. It exploits the changes in the resistance of the fabricating material to temperature changes. Thermistors are passive devices as they need an external voltage supply to operate.

(c) Thermocouple: This sensor is the most commonly used temperature sensor due to its simplicity of manufacturing and precision. A thermocouple consists two dissimilar materials welded together to form 'hot' and 'cold' junctions, which is used to measure temperature by its effect of the development of a voltage difference upon producing temperature difference between these two different materials.

8.3.2 Mechanical sensors

This class of sensors is sensitive to changes in mechanical properties. These sensors are therefore designed to exploit changes to their mechanical properties for quantifying the physical phenomenon. We list some of the common types of mechanical sensors as follows:

(a) Pressure sensor: it is used for measuring the pressure changes in gases or liquids. Pressure sensors can be of various types such as piezoresistive, piezoelectric, capacitive, electromagnetic, strain gauge, resonant, and others [140] (Fig. 8.6).

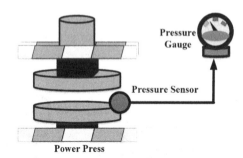

FIGURE 8.6: A typical mechanical sensor arrangement used in detecting pressure changes in an industrial power press.

(b) Barometer: it is kept stable on the ground and is used for measuring atmospheric pressure.

(c) Altimeter: it is operationally similar to a barometer, but measures the changes in the altitude of an object above a fixed level on the ground.

(d) Liquid flow sensor: it measures the liquid flow rate within a bounded channel.

(e) Gas flow sensor: it measures the velocity or flow rate of a gas in a bounded channel.

(f) Accelerometer: it is an electromechanical device and measures the acceleration of a body. Typically accelerometers are either piezoelectric or capacitive.

8.3.3   Electrical sensors

This class of sensors measures and quantifies the electrical properties of objects or their surroundings. We list the following electric sensors based on the particular electrical property they measure:

(a) Ohmmeter: it measures the resistance or changes in resistance of conductive material across its two ends. Principally, it measures the resistance provided by an object to the flow of current through it (Fig. 8.7).

(b) Voltmeter: it measures the potential drop or voltage across a material or device in an electrical circuit.

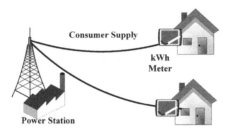

FIGURE 8.7: A typical arrangement showing the positioning of a Watt-hour meter in a consumer power supply.

(c) Galvanometer: it measures current in a circuit by producing deflections in a coil surrounded by a magnetic field. The amount of deflection of the coil is proportional to the amount of current flowing through it.

(d) Watt-hour meter: it is commonly found at the consumer-side terminating junction of consumer electricity supplies. It measures the amount of electrical energy supplied to a device/machinery/household in a fixed duration of time (Fig. 8.7).

8.3.4 Chemical sensors

This class of sensors detect the presence and estimate the amount of various chemical classes in the environment they are set up for monitoring. These sensors use resistive, capacitive, chemical, or biomaterial compounds to quantify the changes produced by an analyte (the substance being analyzed) [141]. The most commonly found chemical sensors include smoke detectors, Oxygen sensors, Carbon Dioxide sensors, Carbon Dioxide sensors, Methane sensors, Butane sensors, alcohol detectors, Nitrous sensors, and others (Fig. 8.8).

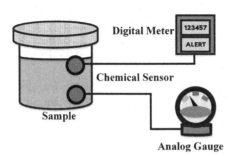

FIGURE 8.8: The variations in chemical sensor types (digital/analog).

8.3.5   Optical sensors

This class of sensors is used for characterizing the optical properties of an object or the environment in general. Generally, these sensors can be resistive or capacitive. These sensors can be used to detect the intensity of light, the frequency of optical wavelengths incident on a particular area (even those lying beyond the visible spectrum) (Fig. 8.9). Some of the common optical sensors are:

(a) Light sensors: these are also known as photodetectors and detects the presence of light.

(b) Photocells: these are also known as photoresistors. They use a variable resistor to determine the changes in light intensity through changes in resistance.

(c) Infra-red sensor: detects infra-red radiation, which lies just beyond the red wavelength in the frequency spectrum of electromagnetic waves.

(d) Ultraviolet sensor: detects ultraviolet radiation, which lies before the violet wavelength in the frequency spectrum of electromagnetic waves.

(e) Color sensor: these sensors typically detect the intensity of the red, green, and blue bands of optical wavelength, which are incident on the sensor (Fig. 8.9).

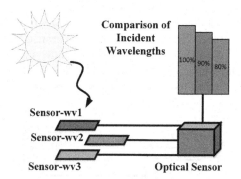

FIGURE 8.9: A typical arrangement showing the detection of various frequencies/wavelengths of light, which is typically found in color sensors.

8.3.6   Acoustic sensors

This class of sensors is responsible for converting the pressure changes due to vibrations from sound waves into electrical energy/signals. The most common example of an acoustic sensor is a microphone, which converts the pressure changes in the surrounding air caused due to various sounds/speech into electrical signals (Fig. 8.10). Other acoustic sensors are seismometers (measures seismic waves to detect earthquakes), and ultrasonic sensors (detects ultrasonic waves and converts them to electrical signals).

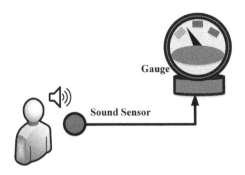

FIGURE 8.10: A typical arrangement for a sound sensor. Here, a microphone detects audible sounds and shows it as a deflection in the meter connected to the microphone.

Summary

This chapter covered the basics of sensing and various sensor types in order to give the readers an idea about the first stage of the industrial automation task. This will help the readers appreciate the importance of sensors in various applications, be it day to day tasks or industrial applications. Readers will also be able to explore innumerable opportunities and ideas arising out of the use of various sensor types, which can be used across a wide variety of domains.

Exercises

1. Differentiate between sensor resolution and its accuracy.

2. What are the various categories of thermal sensors?

3. What happens if a sensor is operated outside its permissible rating?

4. What are the various types of gas sensors?

5. What are the various types of optical sensors?

6. How do acoustic sensors convert sound to energy/electrical signals?

7. How are resistive temperature sensors different from thermocouples?

9

Actuators

Learning Outcomes

■ This chapter provides an introduction to the various actuator types commonly found in industrial environments.

■ New readers will be able to have a grasp of the terminologies and the operating principles of these actuators.

■ We also highlight the various commonly found subtypes of the individual actuator types.

■ The takeaways for each of the actuator types are also highlighted.

9.1 Introduction

This chapter discusses the various actuator types, which are available for use in industrial applications. We discuss the most commonly used actuator types, which will provide the readers with a starting point to pursue selecting actuators for their specific tasks, aimed mainly for industrial scenarios.

An actuator is a device, which converts input signals into physical outputs or mechanical motion. Fig. 9.1 represents the actuation principle. Generally, the input signals are in the form of electrical signals or thermal set-points. Typically, in scenarios such as IoT and IIoT, a processor/controller is responsible for sending electrical signal patterns to the actuators connected to them. The actuators produce desired mechanical motion according to the pattern of the input signals received by them.

Some common examples of actuators include solenoid valves, motors, and pumps. As an example, consider a solenoid valve is used to control the flow of liquid through a pipe. A solenoid valve is an electromagnetic actuator [142]. Whenever an electric current is passed through it, a solenoid coil is magnetized, which results in a proportionate motion of the metal element attached

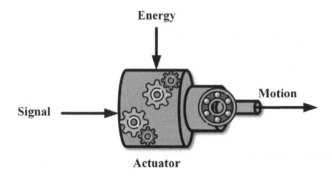

FIGURE 9.1: Basic principle of actuation.

to it. Under normal conditions, when the solenoid valve receives no signal from the processor, the flow through the pipe is unrestricted. As soon as a signal is passed through the solenoid valve, it reduces the flow cross-section of the pipe. This restricts the flow of the liquid through it. Some other classes of actuators, which are commonly found in industrial environments, include thermal, hydraulic, and pneumatic actuators. The various types of actuators are discussed in the subsequent sections of this chapter.

9.2 Thermal Actuators

Thermal actuators operate by converting thermal energy into kinetic energy. These actuators can be considered as a combination of sensing and actuation in a single package. The thermal actuators use temperature sensing to generate responses, which are compared against a fixed set-point. Once the set-point is achieved, the actuation part initiates mechanical motion, which varies depending on the actuator [143]. Applications such as cooling control, freeze control, and boiling point control are typically associated with these class of actuators. Fig. 9.2 shows the working of a wax-based thermal actuator [144]. The wax contained in the bottom chamber expands upon heating, which moves the attached piston upward. The piston comes down as the wax cools. Devices such as water coolers, freezers, boilers, thermostatic balancing valves, and tempering valves are some of the common examples of the use of this class of actuators. Some common types of thermal actuators include:

(a) Thermostatic actuator: These class of thermal actuators relies on a temperature set-point for their operation. Generally, used as control valves, which open or close depending on the temperature difference with respect to the thermal set-point for these actuators. These actuators can be found

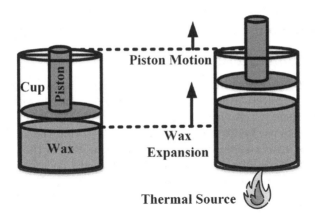

FIGURE 9.2: A typical wax-based thermal actuator in action.

in boiler controls, room heating controls, and other such applications.

(b) Wax motor: These class of thermal actuators typically relies on a source of heat (positive temperature coefficient thermistors) for their operation. Commonly, a wax-like substance is used in this actuator, which expands or contracts in response to applied thermal differences. This expansion or contraction results in the generation of motion of the designed moving mechanism (maybe, a piston).

9.2.1 Takeaways

Thermal actuators have the following takeaways:

(a) They are resilient to power failures as they do not need an active power supply for their operations.

(b) The combination of sensing and actuation in a single package often improves the ease of deployment and reduces unnecessary components in a deployment.

(c) The need for calibrations is removed.

(d) Thermal actuators have a significant hysteresis lag due to the slow speed of moving piston as compared to the speed of change in temperature.

9.3 Hydraulic Actuators

Hydraulic actuators operate by exploiting the compressive property of enclosed fluids to convert it into motion. In simpler terms, these actuators convert hydraulic energy into mechanical energy [145], [146]. Industry-grade hydraulic actuators use highly incompressible fluids so that the applied pressure on the fluid is instantaneously converted to the motion of the moving piston attached to it. Hydraulic actuators work on Pascal's law, which states that due to the law of conservation of energy, the pressure applied to an enclosed fluid is transmitted equally in all directions.

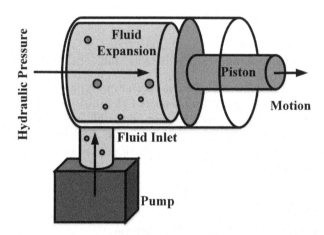

FIGURE 9.3: A typical fluid-based actuator (hydraulic actuator) in action.

Hydraulic actuators can be classified as – (1) linear and (2) rotary. Linear hydraulic actuators produce linear motion upon application of pressure on the enclosed fluid. Some of the commonly available linear hydraulic actuators used in industries are single-acting cylinder types, double-acting cylinder types, and double rod cylinder type. Fig. 9.3 shows a single-acting linear hydraulic actuator. In contrast, the rotary hydraulic actuators produce rotary motion upon the application of pressure on the enclosed fluid. Some of the commonly used rotary hydraulic actuators in industries are rotary-vane type actuator, and rack and pinion type actuator.

9.3.1 Takeaways

Hydraulic actuators have the following takeaways:

(a) They provide a high power-to-weight ratio, which enables these to lift or move heavy weights with the application of significantly low power.

(b) They are suitable for use in situations that require high speed of operations, and the speed is prone to rapid changes.

(c) These actuators are resilient to stall and overload changes. They can resume operation as soon as the overload condition is rectified.

(d) These actuators require careful maintenance and periodic checks against fluid leaks for maintaining hydraulic pressure.

9.4 Pneumatic Actuators

Pneumatic actuators use compression and decompression of air for producing linear or rotary motion [147]. Under industrial settings, these actuators are typically found in valves, which control the flow of liquids or gases. Pressurized air is forced into a chamber to which an air-tight piston is attached. The air moves the piston, which is further used to lift or move a load or drive valves. Typically, a spring-based mechanism returns the piston to its original position once the pressure has been released from the chamber of the pneumatic actuator.

Pneumatic actuators are generally divided into two classes, depending on the return mechanism of the piston. These are:

(a) Single-acting actuators: These class of pneumatic actuators has a single port for letting-in/letting-out pressurized air into the chamber housing the piston. The pressurized air may move the piston forward if the mechanism relies on pushing the piston, or it may pull the piston backward if the machine is designed to suck out air from the chamber. In both cases, a spring-based mechanism is responsible for repositioning the piston to its original location once the effect of pressurized air no longer remains in the chamber. Fig. 9.4 shows an air-based pneumatic actuator. The influx of compressed air pushes the piston forward. The attached spring mechanism is responsible for retracting the piston to its original position.

(b) Double-acting actuators: These class of pneumatic actuators does not have a spring-based piston repositioning mechanism. Instead, they have two air inlet ports, one on either side of the piston. The piston is positioned according to the air burst from the appropriate port. Letting-in pressurized air at the bottom port moves the piston up, whereas putting-in pressurized air from the top port moves it down.

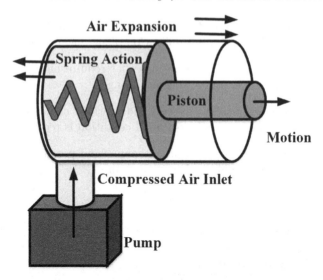

FIGURE 9.4: A typical air-based actuator (pneumatic actuator) in action.

9.4.1 Takeaways

The following are the takeaways from the pneumatic actuator:

(a) These actuators are simple to design, maintain, and easy to use. They are safe to use in environments with high fire hazard chances.

(b) The absence of leaking fluids makes their maintenance significantly easy. This factor also prevents the requirements of cleaning the surrounding machines and systems.

(c) The lower compressibility of air (in comparison to fluids) makes these actuators less efficient.

(d) The cost of repeated use/requirement of compressed air for operating these actuators makes them costlier in the long run.

9.5 Electromechanical Actuators

Electromechanical actuators convert the rotary motion of electric motors to linear motion [148] of shafts through gears. These class of actuators relies on the principle of an inclined plane for their operation. The threads of a lead screw attached to the motor shaft act as a perpetual ramp. The application of rotary motion to the lead screw results in linear motion, which is used for moving loads to large distances. A variety of motors can be found to be in use with this class of actuators depending on the requirements of the task – DC motors, induction motors, brushless motors, and others. Gears are used to reduce the speed of the motors and couple them to their loads. The gears also act to magnify the torque on the loads on which the motors are acting.

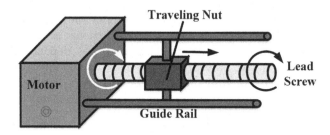

FIGURE 9.5: A traveling-nut type electromechanical actuator in action.

Electromechanical actuators can be of many types. Most commonly, the traveling-nut type and the traveling-screw type actuators are most commonly encountered. The operating mechanism of these two types of actuators are as follows:

(a) Traveling-nut type linear actuator: The motor is attached to one of the ends of the leading screw. The nut travels up and down the length of the lead screw. Any mechanism attached to the nut consequently moves due to the motion of the nut. Fig. 9.5 shows this type of actuator.

(b) Traveling-screw type linear actuator: The motor is attached to the nut. The lead screw passes through the nut. The rotary motion of the motor results in the motor moving up and down along the length of the lead screw.

9.5.1 Takeaways

The following are the takeaways from electromechanical actuators:

(a) This class of actuators has a higher speed of operation and efficiency.

(b) Electromechanical actuators have the possibility of speed control, which helps in a variety of precision tasks.

(c) The parts of this class of actuators are prone to wear and tear, and even burnouts due to overload.

(d) These actuators require electrical power to operate and the operations stop upon the loss of the appropriate driving power.

Summary

This chapter covered the basic principles and types of actuators commonly found in industrial environments. We divide the actuators into four different classes based on their underlying operational principle. Each of these classes is further divided into their sub-types, depending on their means of operation. We only present the most commonly available actuator types, which would enable a reader to find and choose the most appropriate type of actuator depending on the requirements of their applications.

Exercises

1. How are actuators different from sensors?

2. How does a thermal actuator operate? What are the various mechanisms for thermal actuation?

3. How is a hydraulic actuator different from pneumatic actuators?

4. What are the different types of pneumatic actuators?

5. What are the different types of hydraulic actuators?

6. What is the use of gears in electromechanical actuators?

7. What is the difference between a traveling-nut type and a traveling-screw type linear actuator?

8. Where are hydraulic actuators typically used?

10

Industrial Data Transmission

Learning Outcomes

- This chapter provides an introduction to the various data transmission technologies in industrial environments.

- We outline the technology, enumerate its most significant features, and describe the components of these technologies.

- Readers will be able to have a grasp of the terminologies, technologies, and requirements, which are associated with these technologies.

10.1 Introduction

Industrial data transmission is one of the significant aspects of enabling industrial IoT in legacy industrial setups as well as for deciding the structure and architecture of future plants and industrial complexes. Traditionally, industries have relied on field buses for communicating between various controllers and controlled devices. The rise in industrial automation necessitates the need for robust, resilient, and economic measures, which can be easily integrated with existing setups consisting of field devices, controllers, automation servers, and pieces of equipment. Industrial processes are not always limited to the plant floor and may also include requirements for stock monitoring, warehouse controls, building automation, and others. In this chapter, we explore the various data transmission technologies, which are promising for the implementation of Industry 4.0 standards. The data transmission in industries, unlike typical IoT application areas, is often marked with ingenuity, such as using power lines for communications as well as supplying power to devices. Other modes of data transmission in industries include various wired as well as wireless technologies, which we discuss in this chapter.

10.2 Foundation Fieldbus

The Foundation Fieldbus (FF) was developed as a replacement for industrial analog connections by the FieldComm Group (previously known as the old Fieldbus Foundation). This technology uses peer-to-peer (P2P) communication between field devices [149]. The P2P connection enables field devices connected to this network to talk to each other without requiring a master to initiate the connection. This technology easily facilitates services such as field-based controls, diagnostics, and measurement exchanges. Fig. 10.1 shows the architecture and components of a FF. This has been widely adopted by industries such as petrochemical, refining, and nuclear. In extreme cases, temporary devices such as tablets can be allowed to act as the master for this network. FF configures devices and other controllers through the use of function blocks and is majorly of two types based on variations in physical media characteristics, and data rates – (1) H1 and (2) High-Speed Ethernet (HSE).

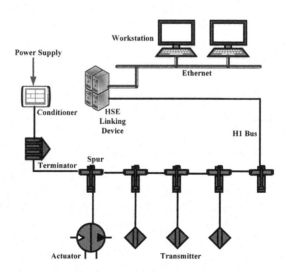

FIGURE 10.1: The FF architecture.

10.2.1 Features

The following are the highlighting features of the FF technology:

- Two or more devices can communicate with each other using P2P networking. Up to 32 devices can be accommodated without the presence of power loops and intrinsic safety features.

- Up to 16 devices can communicate in a P2P manner during intrinsically safe, looped power applications.

- Using only looped power, the number of devices can vary between 1-24, but without any intrinsic safety.

- Typical baud rates of 31.25 kbps are associated with this network.

- FF H1 is used for scenarios that do not require high-speed communication. Communication ensues through Manchester coding.

- FF H1 sends signals alongside voltage using the same set of wires.

- No more than four repeaters can be present in a FF network.

- A maximum segment length of 1900 meters (using type-A cables) is allowed in this network.

- Additional features such as communication failure recovery time of less than 1 millisecond, continuous operations irrespective of devices joining or leaving the network, device insensitivity to polarity inversion, characterize this network technology.

- Device powering should be performed while considering the appropriate polarity of power sources. Typically, each device requires 9 volts to operate, which is supplied through a power supply running parallel to the trunk of FF. In case of the need for additional energy arises for some devices, there are provisions to install extra power supplies with the devices themselves.

- Network operation optimization is done through software, which can calculate the requisite power, current, and voltage requirements depending on the topology being used. The network power consumption is kept lower than the supply power.

Various network topologies supported by this technology include point-to-point (or daisy-chaining), Bus with spurs, Tree, Mixed, and End-to-End. Each of these topologies is briefly enumerated as follows:

(a) **Point-to-point**: The devices are connected in series using their terminals. This configuration is rarely used owing to various operational and maintenance complications for large-scale deployments.

(b) **Bus with spurs**: Devices and spurs are connected to a single bus. Multiple devices may also be connected to each connecting spur, which also dictates the length and number of spurs.

(c) **Tree**: Similar to a star topology, where the trunk of the Fieldbus connects to multiple device couplers and junction boxes. Each of these couplers and junction boxes may further have devices connected to them.

(d) **Mixed**: it is typically a mix of some or all of the possible topologies supported by this Fieldbus. Maximal segment length is restricted by the number of devices, the current flowing through the segment, voltage drop, and spur lengths.

(e) **End-to-end**: This topology enables a direct connection between a field device and other field devices or an H1 card (which is the final control element).

10.2.2 Components

The FF has several components, whose functionalities are itemized below:

- **H1 cards**: These are the Foundation Fieldbus interfaces for H1-based control systems. These connect the network at the controller or the input/output devices.

- **Power supply**: Required for powering the Fieldbus as both signals and voltages are carried over this bus. Several H1 segments can be powered.

- **Signal conditioner**: Enables conditioning of signals before transmitting them over the Fieldbus. This unit provides signal protection from stray impedances through fixed inductances between field devices and power supplies, which also enables good and stable signals to be communicated over the Fieldbus.

- **Wiring**: Typically, shielded twisted-pair are used as the wiring for this Fieldbus, which prevents undue signal reflections and maximizes network length.

- **Segment terminator**: These are used at the ends of trunks to terminate a segment through an impedance matched network properly. Improper segment terminations lead to unnecessary signal reflections and noise during signal transmission.

- **Device coupler**: These are similar to junction boxes, which enable connection to spurs, act as protection for the circuitry, and are used for maintenance and diagnostics.

10.3 Profibus

The Profibus network standard is designed primarily for building and manu-facturing processes [150]. This standard is open and vendor-independent. The Profibus standard boasts of more than 23 million nodes installed worldwide in various industries and locations. The role of Profibus is vital for indus-trial environments in ensuring the robust transition to automation processes from manual ones. As the typical industrial communications need transparent mechanisms for communications, which can proliferate along various organi-zational verticals as well as the horizontal too. Industrial processes rely on robust network standards such as Ethernet, Profibus, and AS-interface due to their usability with various processes as well as lower operational costs.

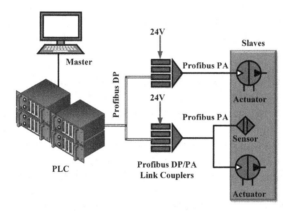

FIGURE 10.2: The Profibus architecture.

10.3.1 Features

The following are the highlighting features of the Profibus technology:

(a) Profibus uses the ASi (Actuator/Sensor interface) for data communication in a cyclic, fast, and efficient manner. The ASi uses a simple and low-cost bus for transmitting data along with the driving circuit voltage (24 VDC) required for operating the sensors and actuators.

(b) Profibus DP or PA variants are used for field-level operations involving I/O devices, transducers, and valves. For the communication of process data, the transmission is cyclic, whereas, for critical signals and alarms, it is done on-demand.

(c) The power and data transmission occur using the same cables (shielded twisted pair).

(d) The PROFINet protocol handles data transmission of large packets between PLCs and PCs, as well as enable integration with the Internet and Intranet through Ethernet.

(e) The addition or removal of devices from the bus does not hinder the configuration or operation of other devices connected to the same bus.

(f) All Profibus variants use the OSI model for communication and rely on the ISO 7498 international standard. It uses Layer 1, 2, and 7 only for efficiently meeting the requirements of communications, using the FMS.

10.3.2   Components

The Profibus architecture is differentiated into three variants – Profibus DP, Profibus FMS, and Profibus PA. Fig. 10.2 shows the Profibus architecture and its components. We highlight the defining aspects of these Profibus variants as follows:

- Profibus DP: It is designed for high-speed communications between automation systems and pieces of equipment. The communication is enabled using fiber optic cables or RS-485 physical medium. Fiber optic cables enable transmission over 80 km range. RS-485 uses NRZ asynchronous coding for transmission of data through two configurations of 9.6 kbit/s and 12 Mbit/s. It typically uses twisted pair shielded cables, and the transmission distance ranges from 100 m for 12 Mbits/s to 1 km for 9.6 kbit/s. The bus configuration of RS-485 allows for devices to be added or removed from the network without affecting other devices. This standard is designed for time-critical communication for control and requires approximately 2 minutes for transmitting data of 1 kilobyte from Input/Output (I/O) devices. This variant of Profibus is the most widely used in applications relying on master-slave architecture and is available in three versions – DP-V0 (1993), DP-V1 (1997), and DP-V2 (2002).

- Profibus FMS: it is primarily designed for P2P communication tasks between Programmable Logic Controller (PLC) and Distributed Control Systems (DCS). This universal standard provides a range of functions for achieving complex communication tasks, not limited to data exchange between intelligent systems.

- Profibus PA: it is designed for enabling communication between process automation and control systems and field equipment such as sensors, converters, and actuators. This standard ensures high-speed controls, reliability of transmission, increased range, increased economic benefits, and reduced operational costs. It uses Manchester encoding for message transmission at a rate of 31.25 kbit/s in voltage mode. The enhanced safety features correspond to the guidelines laid for critical infrastructures such as those found in chemical industries.

10.4   HART

The Highway Addressable Remote Transducer or HART was designed to provide point-to-point digital communication between field devices and a central control system [151]. This protocol uses a master-slave architecture. The slave devices do not elicit any response until instructed by the master device.

10.4.1   Features

The HART protocol has the following highlighting features:

(a) This protocol uses Frequency Shift Keying (FSK) for modulating information onto the 4–20 mA power signal for enabling digital communication.

(b) HART enables a bidirectional mode of communication between smart field devices.

(c) Process variables, as well as additional information from the smart field devices, can be communicated to a monitoring or control device.

(d) It does not require special wiring or terminations and is used over standard instrumentation grade wires.

(e) HART provides a data rate of 1200 bps over the 4–20 mA signal.

(f) HART master devices or applications are updated regarding the status of its connected field devices at least twice per second.

(g) It has two communication channels – (1) 4–20 mA industrial standard for reliable communication of the measured analog data, and (2) digital signaling through FSK over the analog channel for communicating additional device information from the field devices.

(h) The digital channel carries information about device diagnostics, status, and other such information.

(i) HART is considered a very low-cost, robust, and easy to configure industry standard.

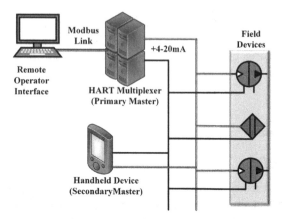

FIGURE 10.3: The HART architecture.

10.4.2 Components

The HART protocol follows a master-slave communication and control approach. This configuration allows the protocol to communicate in a point-to-point or multidrop fashion between the smart field devices and the central control and monitoring system. A burst mode of communication, which is only supported by the point-to-point configuration, allows for a higher data-rate continuous broadcast of HART reply messages from a single slave device. Up to two masters is supported by HART, without interfering with the communication. The first master or the primary master status is designated to the central control and monitoring system. Handheld communicators are designated as secondary masters. Fig. 10.3 shows the HART architecture and components.

10.5 Interbus

Interbus is a fast and robust enterprise-grade networking protocol relying on serial data transmission for communication between controllers, systems/computers, and sensors/actuators [152]. It follows a ring topology for connecting all the devices in its network and has a master-slave architecture. The main advantage of this protocol is its low susceptibility. More than 17 million devices world over, use this protocol. Modern systems can connect Interbus with Ethernet for achieving standardization of industrial communication. At any time, Interbus supports a maximum of 512 subscribers to its network, which can leave or join the network without hindering the operation of the other subscribers.

10.5.1 Features

The following are the highlighting features of Interbus technology:

(a) Interbus provides a data rate of 500 kbps.

(b) The presence of branch bus terminals prevents the spread of faults from a sub-network to all over the network.

(c) The main remote bus is branched at various locations through bus terminals to form local branches.

(d) Individual local branches can be switched on/off independent of each other.

(e) Interbus enables automatic installation, diagnosis, easy maintenance, and operation of its network through automatic addressing.

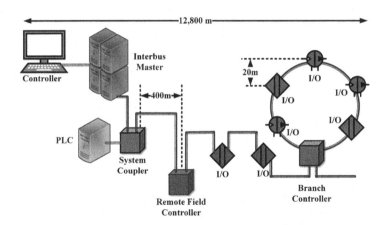

FIGURE 10.4: The Interbus architecture.

10.5.2 Components

The remote bus is primarily responsible for communication as well as the power supply in the Interbus network. However, its range between two devices is limited to 400 m at most. Branch networks connect to the main Interbus network through branch terminals. Further branching of branched networks is not allowed in Interbus. The Interbus loop, which connects the sensors and actuators in a ring topology, has a maximum range of 20 m between two devices/sensors/actuators. The elements of the ring function as slaves, which are controlled by a master. The master connects to the remote bus. Fig. 10.4

shows the Interbus architecture and its components. The master's connection with the loop is known as the connection group. The CDM or Configuration Monitoring Diagnostic is another useful aspect of Interbus. CDM is typically used during the process planning stage to configure the program addresses of all the elements of this network. CDM is also able to maintain and operate the connection group. It also helps in sorting faults in the network, such as short circuit faults, transmission pauses, and others. The Interbus protocol also facilitates the identification of wear and tear of network elements through statistical fault evaluation.

10.6   Bitbus

The Bitbus is a non-proprietary and open Fieldbus technology [153]. Intel introduced it as a Fieldbus component for industrial communication. It is made up of two core components – (1) an RS-485 interface, and (2) the software side. The RS-485 interface handles the physical connectivity aspects of this technology, whereas the software side relies on Synchronous Data Link Control (SDLC) for serially transmitting data in a transparent manner. Both of these components are individual standards. The Bitbus ensures low total cost of ownership, cross-compatibility with the various OS, and high reliability, among others. It also enables the field delegation of control in a real-time manner.

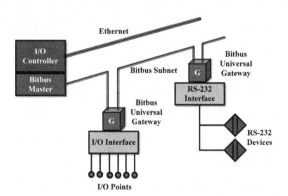

FIGURE 10.5: The Bitbus architecture.

10.6.1 Features

The following are the highlighting features of Bitbus:

(a) It has two different data encoding modes – Synchronous and Self-clock modes.

(b) The synchronous mode of transmission requires an additional pair of wires for enabling communication, which is marked by the transmission of a synchronous impulse.

(c) The self-clock mode removes the use of common repeaters and level converters as it uses the Non-Return to Zero Inverted (NRZI) encoding for data transmission.

(d) Each Bitbus segment in a bus topology can accommodate at most 28 devices. This number is typically enhanced to 250 using Bitbus-compatible repeaters.

(e) The Bitbus lines can be either 300 m or 1200 m long.

(f) The Bitbus supports a data rate of 375 kbps for a 300 m line or 62.4 kbps for a 1200 m line.

(g) The use of repeaters significantly reduces the data rate. For example, the data rate can fall to 62.4 kbps just by using two repeaters.

(h) In case of applications requiring low data rates, a maximum of ten coupled repeaters can be supported. Beyond this number, the losses become significant and start to hamper the reliability of Bitbus.

(i) The use of coupling repeaters require the use of 120 ohm resistance connections at each end of the line.

(j) The number of addresses supported by Bitbus is 256. Each device on the Bitbus has an address indicated by the number ranging from 1 to 249. The remaining numbers cannot be assigned as regular addresses.

10.6.2 Components

The Bitbus data transmission speed is proportional to the length of the bus connection. A typical Bitbus length of 300 m has a data rate of 375 kbps, which gets drastically reduced to 62.4 kbps if the bus length increases to 1200 m (four times increment). The Bitbus uses RS-485 specifications to define its physical media (wiring). It primarily relies on the twisted pair wiring for operations, with the signal ground acting as the shielding from interference. Fig. 10.5 shows the Bitbus architecture and its components. The Bitbus plug assignment is as follows

- Pin-1: not assigned

- Pin-2: not assigned

- Pin-3: DATA B (−)

- Pin-4: RTS B (−)

- Pin-5: Ground

- Pin-6: not assigned

- Pin-7: not assigned

- Pin-8: DATA A (+)

- Pin-9: RTS A (+)

10.7   CC-Link

Mitsubishi Corp developed CC-Link, which stands for the Control and Communication Link [154]. Developed initially as Mitsubishi's Fieldbus solution for their internal use, CC-Link has now become a popular open network for automation and communication as it has made its data transmission protocols available for integration with various interfaces and developments. Currently, there are four versions of CC-Link – (1) the standard CC-Link, (2) CC-Link Safety, (3) CC-Link LT, and (4) CC-Link IE. CC-Link enables both high-speed data as well as control sequences, which makes it quite robust for use in process automation tasks. The most significant advantage of this technology is its network range of about 14 km. The CC-Link Safety is a Fieldbus for ensuring a safe system architecture. It is tasked with identifying fault-generating errors, after which the systems connected to it are brought to a safe state. Another variant of CC-Link, which was designed for use with the connection of sensors and actuators with the primary circuit, is the CC-Link LT. It is a bit-oriented system. Finally, the CC-Link IE variant ensures error-free data communication of large data through the Ethernet. The IE stands for industrial Ethernet. It enables the use of existing networks for connecting various devices in an optimized manner. The CC-Link IE is further divided into two types – (1) CC-Link IE Control and (2) CC-Link IE Field. Fig. 10.6 shows the CC-Link architecture and its components.

10.7.1   Features

We briefly outline the highlighting features of the variants of CC-Link technology as follows:

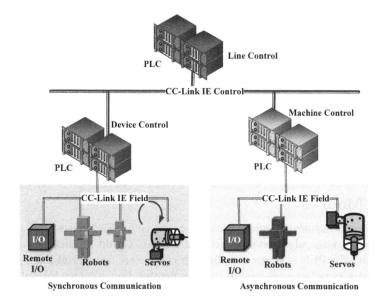

FIGURE 10.6: The CC-Link architecture.

- Standard CC-Link: The basic CC-Link technology achieves a network transmission rate of 10 Mbps, uses a master-slave architecture, and uses RS-485 for defining the transmission standards. A distance of 1200 m is characteristic of this variant. It can be extended to 13200 m through the use of repeaters. However, it reduces the network refresh time. Each network can support up to 64 stations.

- CC-Link Safety: It is similar to the basic version of the CC-Link. However, it adds the benefits of the integration of safety devices with regular devices on the same network.

- CC-Link LT: It achieves a transmission rate of 2.5 Mbps and supports 64 stations in a network. It uses a trunk length of 800 m, which includes a drop length of 200 m.

- CC-Link IE:

  - CC-Link IE Control: It is deterministic and uses a 1 Gbps optical fiber and Ethernet-based network. It supports 120 stations in a network, following a master-slave architecture, with a maximum distance of 550 m between stations.

  - CC-Link IE Field: It is also deterministic and uses a 1 Gbps Ethernet-based network. It supports 121 stations per network with a maximum distance of 100 m between stations.

10.7.2 Components

The CC-Link network comprises two broad components – master and slaves. The master of a network enables communication between the various slaves connected to its network. Typically, PCs and PLCs act as network masters, whereas devices such as valves, sensors, actuators, robots, and drives act as networked slaves.

10.8 Modbus

Modbus is an industrial communication protocol for Fieldbuses [155]. It is one of the most commonly used industrial field buses. Gould-Modicon developed it as a means for PLC systems to talk with computers over serial links. It is one of the most popularly associated communication technologies for automation systems over the Ethernet. As the Modbus does not rely on the transmission medium, it acts as a universal communication protocol, which can accommodate a massive variety of industrial systems. Modbus follows a master-slave architecture. Fig. 10.7 shows the Modbus architecture.

10.8.1 Features

The following are the highlights of the Modbus features:

- It was designed for transmitting data over serial links.

- It supports a master node and 247 slave devices, where each slave device is allocated a unique address from 1 to 247.

- The data is transmitted in the form of bits over a single cable connecting two devices.

- A Modbus map defines the type of data from each sensor (temperature, humidity, and others), the location of its storage (addresses), and the nature of the data (byte, string, and others).

- Modbus can transfer data through three modes – (1) TCP, (2) RTU, and (3) ASCII.

10.8.2 Components

The Modbus TCP is designed primarily for use with Ethernet. Each message starts with a 2 Byte long sequence number and is truncated by the message length. It uses TCP port 502 for communication. Its lower layers ensure error

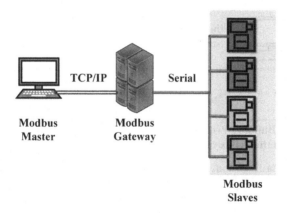

FIGURE 10.7: The Modbus architecture.

protection and do not require the calculation of the checksum. The Modbus Remote Terminal Unit (RTU) transmits data in the form of binary codes and was designed for serial interfaces. RTU messages start and end with a 3.5 character long pause message. The ASCII data starts with a colon (:) and the receiver address. The ASCII message structure ends with a CRLF. It is designed to make the data human-readable.

10.9 Batibus

Batibus was initially developed as an enhancement of typical industrial communication systems to cater to complex building automation tasks [156]. Currently redundant, Batibus was replaced by the Konnex standard. However, it is accredited with being one of the primary open Fieldbus protocols for buildings and apartments by enabling communication between sensors/actuators and other field devices. Batibus uses a twisted pair cable system for communication between sensors and actuators. Applications such as temperature control, humidity control, ventilation systems, alarms, and security were some of the ones enabled by this protocol. Fig. 10.8 shows the Batibus architecture and its components.

10.9.1 Features

The following are the highlighting features of Batibus:

(a) It is an open Fieldbus protocol.

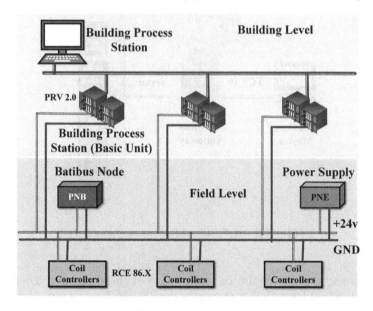

FIGURE 10.8: The Batibus architecture.

(b) It uses twisted-pair cabling for communication, which makes the topology simple and economical to implement.

(c) Each Batibus station can be identified by their individual addresses.

(d) Batibus stations can be added or removed easily from the network topology without affecting the functioning of the network.

(e) Batibus has been standardized as JTC1/SC25 under ISO/IEC standards.

10.9.2 Components

Batibus is divided into two components, which highlight its operations – (1) access procedure and (2) cables and topologies. Batibus relies on CSMA/CA for communication over its network. CSMA/CA is necessary as all the stations are connected to the same bus and need robust solutions for medium access. The cables used in Batibus are twisted pairs. Using this simple form of cabling topologies such as a tree, star, ring, and line are supported. Each network node in Batibus can have one address out of 240 available ones. An additional 16 addresses are reserved for group allocation, using which any individual node within a grouping can be accessed. Using the cabling system, Batibus can achieve a maximum data transmission rate of 4.8 kbps. Each network segment in Batibus is supplied with 15 V and 150 mA externally, for enabling its operations.

10.10   DigitalSTROM

The DigitalSTROM is a patented transmission technology for enabling the control and maintenance of electrical systems through the existing power lines themselves [157]. This is enabled by installing a special terminal at the electrical systems, which are controlled through control equipment via a DigitalSTROM chip circuitry.

10.10.1   Features

The following features characterize the DigitalSTROM:

(a) The information between various connected components is transmitted through power lines, which already exist in buildings.

(b) This technology uses the same power lines as forward as well as backward communication channels.

(c) It follows a master-slave architecture.

(d) Data transmission to the master node is achieved by changing the power of the sinusoidal wave of the voltage across the power lines for brief instants of time.

(e) Each of the DigitalSTROM components has a globally unique identification, similar to a MAC address.

(f) The connected network components are assigned local addresses using a mechanism similar to the traditional DHCP.

10.10.2   Components

The DigitalSTROM comprises three parts – (1) the DigitalSTROM meter (dSM), (2) DigitalSTROM Identification Device (dSID), and (3) DigitalSTROM Server (dSS). Fig. 10.9 shows the DigitalSTROM architecture and its components. The dSM is similar to an ammeter and acts as the bus meter. The dSM transmits data on the power lines through the use of the DigitalSTROM protocol, which is hosted on the dSM. dSMs can be hooked onto single or multiple phases for enabling communication. Each dSM has its individual circuit with multiple DigitalSTROM Devices (dSD). The dSID is a slave and has a unique ID in the network. A maximum of 128 slaves can connect simultaneously to a dSM circuit. The dSID uses a serial interface and is coupled with the Line (L) and Neutral (N) of the mains current. When functioning as an actor, one of the outputs of the dSID device is on the L, which is controlled by the DigitalSTROM chip. The dSS connects the various DigitalSTROM units with

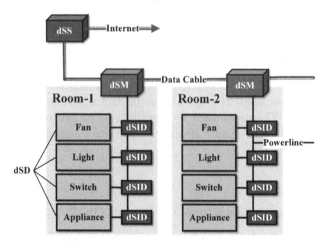

FIGURE 10.9: The DigitalSTROM architecture.

themselves as well as external applications. It requires an operating voltage of 24 V and can be considered as a gateway to which various features can be added. It also acts as the network storage and keeps a record of the temporal changes in its network.

10.11 Controller Area Network

The Controller Area Network (CAN) was developed by Bosch and standardized under ISO11898-1 as a serial communication bus for industrial and automotive applications [158]. CAN was originally designed as a solution for the problem of the requirement of extensive wiring in vehicular communications. CAN provides high-speed communications as well as separates control and data messages being transmitted over a single wire system. Fig. 10.10 shows the CAN architecture in the context of a vehicle.

10.11.1 Features

The following are the highlighting features of the CAN bus protocol:

(a) CAN defines the layer-1 and layer-2 of the OSI model for in-vehicle communications over a single wire system.

(b) It is a CSMA/CD protocol. Upon detection of collisions, the message with the highest priority is transmitted first by a process known as priority arbitration.

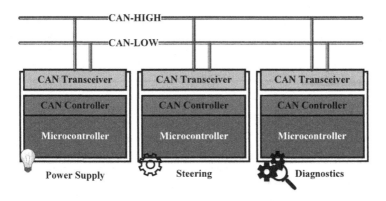

FIGURE 10.10: The CANbus architecture.

(c) Vehicle diagnostics and diagnostic codes are transmitted using the CAN bus.

(d) CAN uses a multi-master hierarchy for accessing the bus as all messages from all nodes are broadcast over the network.

(e) CAN has 2048 distinct message identifiers, which also denote the priority of each message.

(f) It uses an inverted logic for signaling and has a differential signaling state represented by recessive and dominant states.

(g) CAN messages undergo Cyclic Redundancy Checks (CRC), and also relies on ACK messages for indicating successful message transfer.

10.11.2 Components

CAN uses a single wire-based communication for establishing a high-speed in-vehicle network. It relies on four primary message types for its operations – (1) data frame, (2) error frame, (3) overload frame, and (4) remote frame.

A CAN data frame has the following sequence – (1) start of frame bits (SOF), (2) identifier bits (ID), (3) remote transmission request bits (RTR), (4) control bits, (5) data bits, (6) data length bits, (7) CRC bits, (8) acknowledgment bits, and (9) end of frame (EOF) bits. The RTR and ID are dominant, whereas ACK is recessive in the data frames. Upon successful transmission, the ACK is overwritten by a dominant bit.

The CAN error frame results in all the CAN nodes generating error frames until the source of the error retransmits the message.

The overload frame is generated by a node when it receives data at a rate faster than it can process. The overload frame is somewhat similar to the error frame.

The CAN remote frame is sent to a node to request data from it. It is similar to the data frame, but without any data and RTR bit as recessive.

10.12   DeviceNet

DeviceNet was designed by Rockwell Automation to enable communications between PLCs and field devices such as sensors, actuators, and valves. It is designed as an application-level protocol for automation systems. DeviceNet uses the Common Industrial Protocol (CIP) and uses the power line cables for transferring power as well as communicating [159].

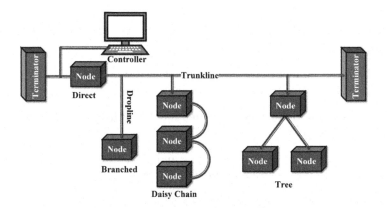

FIGURE 10.11: The DeviceNet architecture.

10.12.1   Features

The following features characterize the DeviceNet protocol:

(a) The DeviceNet physical layer uses a trunkline-dropline topology for communicating with devices. CAN enables the DeviceNet data link layer, whereas the DeviceNet network layer relies on device MAC ID, and the transport layer relies on message IDs for networking.

(b) DeviceNet supports 64 nodes simultaneously. The connected systems are addressed using one of the numbers lying between 0-63.

(c) It uses five cable types, all of the twisted pair type, which is used interchangeably depending on the distances and requirements of the application – thick round, thin round, class-1 round, KwikLink flat, KwikLink lite flat.

(d) It uses 24 V for powering the network components through one of the cables, and the other part of the pair is used for signaling.

(e) DeviceNet is characterized by the data rates of 125 kbps, 250 kbps, and 500 kbps. These data rates decrease depending on the increase in cable length.

(f) The devices are connected to the DeviceNet trunkline through droplines.

10.12.2 Components

The DeviceNet dropline length is limited to approximately 6m, which means that the systems/devices connected to the trunkline are restricted within 6 m of the trunkline. The DeviceNet terminating resistors insulate the DeviceNet from electrical noises and require a 121 Ohm, 1/4 Watt resistors at both ends of the trunkline. Trunklines can be connected in various configurations (daisy-chain, branched, direct) to various devices through taps and connectors. Fig. 10.11 shows the DeviceNet architecture and some of its topologies.

10.13 LonWorks

The LonWorks [160] system is an open network initiative, which consists of both hardware and software components and was developed by Echelon Corp based on and uses the seven layers of the ISO-OSI stack. LON stands for Local Operating Network. The hardware has a core component referred to as the 'Neuron chip.' The software component is known as LonTalk. The LonWorks devices are interoperable and enable communication between different entities such as sensors, actuators, and other control devices. The simplicity of this protocol is that all the devices operation the LonWorks protocol (nodes) are considered as objects, which, when brought together, can establish a complex end-to-end task path such as an assembly line (Fig. 10.12).

10.13.1 Features

The various highlighting features of LonWorks are as follows:

- Network variables and software definitions foster the concept of interoperability in LonWorks. These are known as Standard Network Variable Types (SNVT).

- LonWorks consists of four components – (1) Neuron chips, (2) LonWorks transceivers, (3) LonTalk protocol, and (4) Network Management software.

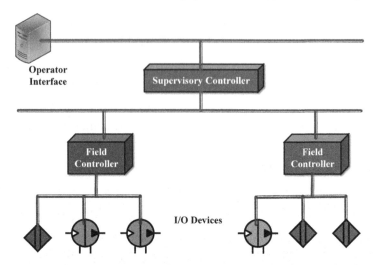

FIGURE 10.12: The LonWorks architecture.

- The physical medium can be accessed using either a wired or wireless communication medium. The wired media supporting LonWorks include powerline cables, coaxial cables, fiber-optic lines.

- It supports a maximum transmission rate of 1.25 Mbps.

- The field controllers are responsible for direct control of the I/O devices.

- Multiple field controllers are commanded by a supervisory controller, to which an operator interface directly connects.

10.13.2   Components

The LonTalk messages are typically prioritized for collision aversion. It uses a predictive collision detection algorithm, which is proprietary. This protocol supports up to 500 transactions per second and uses a node prioritizing mechanism so that the node with the highest priority can access the physical medium as soon as an ongoing transmission is completed. The LonWorks transceivers have the following specifications:

- Twisted pair transceiver: The data rates are either 78 kbps or 1.25 Mbps, with high common-mode noise rejection. A maximum network length of 1.4 km using loop or backbone configuration is achieved for the 78 kbps connection types, whereas a mere 131 m is achievable using the 1.25 Mbps connection type.

- Power line transceiver: The data rate of approximately 3.6 kbps is achievable using this transceiver. However, this medium removes the need for additional wiring for communication as it can use the existing power-lines for communication. A maximum network length of approximately 1.99 km is achievable using this configuration.

- Twisted pair, free topology transceiver: A data rate of 78 kbps is achievable using this transceiver with a network length of approximately 488 m. Using repeaters, this can be increased to approximately 965 m. This transceiver supports almost all network topologies, including bus, star, loop, and mixed configurations.

- Link power twisted pair transceiver: This transceiver is similar to the previous one. However, the power is centrally distributed over the network.

- RF transceiver: This transceiver enables wireless communication and operates on 400 MHz, 470 MHz, and 900 MHz bands.

- LonMaker integration tool: It is a program for designing LonWorks networks and plan them. It also helps in maintaining and installing the LonWorks networks through a Microsoft Visio-based GUI.

10.14 ISA 100.11a

The ISA100.11a standard was developed by the International Society of Automation (ISA) as an open standard for enabling wireless and robust industrial automation [161]. It relies on IPv6 for enabling networking on IP-based devices and uses the IEEE 802.15.4 standard specifications for the physical and data-link layers (bottom two layers of the OSI stack). It is a very low-power protocol but includes security measures, including AES-128 based encryption.

10.14.1 Features

The following features characterize ISA100.11a:

(a) It operates at the 2.4 GHz ISM band, which supports 16 channels.

(b) Interference mitigation is achieved using Frequency Hopping Spread Spectrum (FHSS) modulation. A special feature called "channel blacklisting" is incorporated to reduce noise and interference.

(c) Besides the inclusion of AES-128 encryption, ISA100.11a also ensures security at the data link and transport layers.

(d) It can be integrated into wired or backbone networks through a gateway.

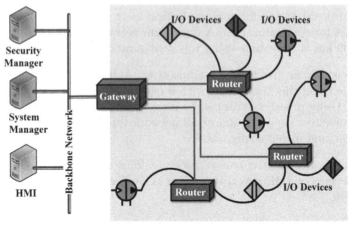

FIGURE 10.13: The ISA100.11a architecture.

(e) TDMA is used for media access and CSMA channel state sensing.

(f) The network layer is based on IPv6, similar to the 6LoWPAN protocol.

(g) The transport layer uses UDP for communication.

(h) The supported network topologies include star and mesh.

10.14.2   Components

A central system manager handles the communication scheduling and network routing in this network. Fig. 10.13 shows the ISA100.11a network architecture. It supports a variety of field devices, including both IP-based as well as non-IP-based ones. The field devices may be mobile, static, or handheld. The field devices connect to backbone devices. The backbone devices are static and include gateways, routers, system, and security managers. All the signals and data from the field devices and routers are forwarded to the gateway, which in turn connects this wireless network to the wired backbone network.

10.15   Wireless HART

The WirelessHART devices are similar to the HART devices but were built to enable wireless communication for HART-based field devices [162]. It is an open and free standard, which was standardized under IEC 62591. It follows a mesh networking topology and is backward compatible with old HART devices. It operates on the 2.4 GHz ISM band and uses Direct Sequence Spread Spectrum (DSSS) with channel hopping over the physical medium.

10.15.1   Features

The following features characterize wireless HART:

(a) Wireless HART physical layer is based on IEEE 802.15.4, the same standard used by protocols such as Zigbee, 6LoWPAN, and Thread.

(b) It operates on the 2.4 GHz band and provides a data rate of 250 kbps.

(c) Channels 11 to 26 of the 2.4 GHz spectrum are used for radio transmission. Adjacent channels have a gap of 5 MHz in between them.

(d) It uses DSSS and Orthogonal Quadrature Phase Shift Keying (OQPSK) for modulating information over the physical media.

(e) TDMA using bus arbitration through the use of multiple timeslots grouped into superframes is employed.

(f) The acknowledgment packets are also responsible for synchronizing timeslots in the Wireless HART network.

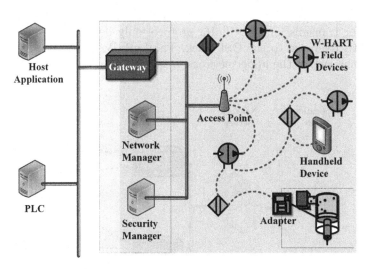

FIGURE 10.14: The WirelessHART architecture.

10.15.2   Components

The WirelessHART network is formed of six main components – (1) field devices, (2) handheld devices, (3) adapters, (4) gateways, (5) network manager, and (6) security manager. Fig. 10.14 shows the Wireless HART architecture and components. The field devices have routing capabilities and are installed for collecting data or actuation. The handheld devices are similar to field devices but are mobile, unlike the field devices. The adapters are devices that can be used to upgrade HART devices to make them wireless. The network manager is responsible for the network functions. The security manager is tasked with handling the security aspects of the formed network. The gateway connects the internal (location-specific) HART network to the backbone network or the plant-wide communication bus. Each gateway supports up to 250 field devices.

10.16   LoRa and LoRaWAN

Long Range or LoRa [163], as it is most commonly known, was developed by Cycleo for communicating in the sub-GHz spectrum over long distances wirelessly. It was later acquired by Semtech to form the LoRA Alliance. It is a reasonably recent Low-Power Wide Area Network (LPWAN) technology for enabling IoT and relies on Chirp Spread Spectrum (CSS) for modulating signals for transmission. The LoRa protocol only defines the lower physical layer with reference to the OSI stack. The LoRaWAN protocol emerged to address the lack of the upper network layers of LoRa.

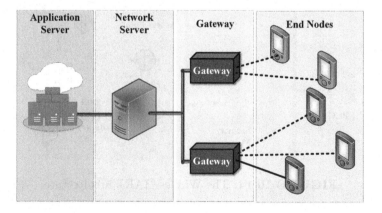

FIGURE 10.15: The LoRaWAN architecture.

10.16.1   Features

The following features characterize LoRa and LoRaWAN:

(a) LoRa operates on the sub-GHz radio frequency spectrum in the bands of 433 MHz, 868 MHz, and 915 MHz.

(b) Transmission ranges of over 10 km are achievable using this protocol.

(c) Modern LoRa devices are characterized by their very low-power consumption and long communication ranges.

(d) LoRa devices can use geolocation and triangulation features. However, these features are not direct and depend on gateway-based timestamps.

(e) LoRa uses a proprietary version of the CSS and also incorporates forward error correction to mitigate interference and noise.

(f) Devices transmit data asynchronously on a need basis.

(g) LoRaWAN manages the communication between end devices and the gateway.

(h) LoRaWAN is a cloud-based MAC protocol but functions as a network protocol.

10.16.2   Components

As LoRa only defines the physical layer of the OSI stack, LoRaWAN emerged to handle the networking aspects of communication. LoRa devices have only the technology for long-range communication, but LoRaWAN defines the various operational communication protocols to enable this long-range communication. LoRaWAN manages device features such as power, data rate, and frequencies. Fig. 10.15 shows the LoRaWAN architecture. LoRa end-devices transmit data to multiple gateways, which aggregate the data and forward it to a central network server. The central server handles tasks such as removing packet redundancies, security, and network management. Network servers further forward the data to application servers. The application servers are typically hosted at the cloud.

10.17   Recent and Upcoming Technologies

Some of the recent and upcoming technologies, which are rapidly being adopted for industrial uses as enabling technologies for IIoT, are NB-IoT and IEEE 802.11 AH.

10.17.1   NB-IoT

The Narrow Band IoT or NB-IoT is yet another Low Power Wide Area Network (LPWAN) technology to enable robust and power-efficient communication over long distances [164]. Designed mainly for low-bandwidth applications, NB-IoT devices can last more than ten years on a single charge for general applications. It is designed to co-exist with existing cellular technologies such as 3G and 4G. NB-IoT has been designed to be versatile during its deployment as it uses a 200 KHz spectrum for communication. It can exist independently in licensed bands, or LTE base stations can allocate resource blocks or guard bands, or in the unused 200 KHz band (GSM/CDMA). NB-IoT has the capability of providing connectivity indoors and even deep undergrounds. It can also accommodate a massive number of low delay-sensitive devices at very economical costs.

10.17.2   IEEE 802.11AH

The new wireless networking protocol, which provides an improvement over the standard IEEE 802.11 WiFi standard, is the IEEE 802.11AH, which is also known as WiFi HaLow [165]. The IEEE 802.11AH operates on the license-free 900 MHz band for providing enhanced network coverage. Traditional WiFi uses the 2.4 GHz or the 5 GHz RF spectrum. This new protocol's power consumption is significantly low despite the higher data rates and enhanced coverage area. It uses OFDM for modulating signals on the physical medium. Channel widths of different specifications ranging from 1, 2, 4, 8, and 16 MHz are available depending on the range and use-case. Data rates of up to 347 Mbps are achievable using this protocol. Through the use of a compact MAC header format and a new medium access scheme, the IEEE 802.11AH significantly increases its throughput.

Summary

This chapter covered the basics of data transmission technologies in industries and IIoT. We covered the most commonly used field buses and wireless technologies to illuminate the reader about the existing solutions and their scope in various industrial environments. Finally, we also provide a glimpse into the few recent and upcoming connectivity technologies which will broaden the scope and possibilities of using industrial IoT.

Exercises

1. What are the crucial components of a FF?

2. How is Foundation Fieldbus different from ISA100.11A?

3. Differentiate between the Profibus and the Interbus.

4. How does a CAN work?

5. How is CC-Link IE control different from the CC-Link IE Field?

6. What differentiates LonWorks from Devicenet?

7. What are the primary functions of the Neuron chip in the LonWorks protocol?

8. Differentiate between HART and Wireless HART.

9. Differentiate between ISA100.11A and Wireless HART.

10. What is Batibus?

11. How is LoRaWAN different from LoRa?

12. What is the most promising feature of the NB-IoT protocol?

13. What differentiated the IEEE 802.11AH from other wireless standards?

11

Industrial Data Acquisition

Learning Outcomes

- ■ This chapter provides an introduction to the various control infrastructures commonly found in industrial environments.

- ■ New readers will be able to have a grasp of the terminologies, and components of industrial control systems.

- ■ Seasoned learners will be able to appreciate the types of control solutions popularly used for industrial controls and their dependencies.

11.1 Introduction

Industrial control systems are typically characterized by the presence of sophisticated instrumentation devices, hazardous process monitoring requiring real-timeliness, high density of deployed devices, harsh environments, interactive controls, and economic constraints. These control systems acquire data through various deployed sensors, which measure the process variables (PV) and compare them against desired setpoints (SPs) to obtain control functions which are affected through final control elements (FCE) to control the process [166]. Large-scale industrial controls are achieved by systems such as Distributed Control Systems (DCSs), SCADA, and PLCs. We discuss some of these large-scale control systems in industrial environments in this chapter.

11.2   Distributed Control System

The DCS was designed for industrial processes with distributed controllers and large control loops. A centralized supervisory control manages the distributed controllers. DCS supports a wide range of fieldbuses and digital communication buses. The use of DCS enhances the scalability, security, and reliability of plant control processes. Primarily, DCS enables remote supervision and brings the control functions nearer to the plant processes themselves [167], [168]. DCS is especially useful in plant processes that follow batch processing operations such as manufacturing plants, chemical plants, nuclear power plants, and water management plants. Fig. 11.1 shows the typical DCS architecture and its various components.

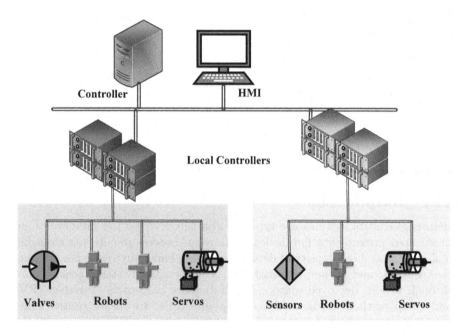

FIGURE 11.1: The DCS architecture and components.

11.2.1   Components

DCS comprises of the following four elementary components:

(a) Controller: It is also referred to as the engineering station. It controls the various distributed controllers or local control units by implementing control algorithms for various devices and performing supervisory control. It also manages the network communication.

(b) Local control unit: It is also referred to as the distributed controller. These are placed near field devices and directly control them over local communication links by evaluating process variables against desired setpoints. The local control units can handle both analog and digital inputs and outputs.

(c) HMI: The human-machine interface or HMI is tasked with logging data from various plant processes and other control-based operations, analyze their behavior through interactive visualizations, infer trends, and project targets. HMIs also serve as alarm systems and can be provided with control functions also.

(d) Communication protocols: These consist of various cabling systems (coaxial cables, optical fibers, and others), along with various communication protocols required for the functioning of the DCS. The DCS architecture uses at least two communication protocols for achieving information transmission – (1) between the field devices and the local control units and (2) between the local control units and the engineering station.

11.3 PLC

A PLC or a Programmable Logic Controller is an industrial-grade computer-based control system for electro-mechanical automation tasks in harsh and rugged operating environments. Industrial processes such as manufacturing plants, chemical plants, and assembly lines rely on PLC-based automation for highly coordinated sequential execution of tasks [167]. PLC provides the features of reliability, timeliness, fault diagnosis features, and is additionally very easy to program, making it very robust for a wide variety of industrial applications [169].

PLCs have been a very convenient replacement for hard-wired systems in industries, as typically, industrial environments have tens of thousands of hard-wired components. Replacing or changing the configuration of hard-wired components is costly as well as time-consuming. The replacement of components may also affect the functioning or performance of its neighboring components, which may share a part of the hard-wired network. In contrast, PLCs can reprogram their configuration [170], making the maintenance and upgrade tasks quite economical, timely, and less laborious.

Functionally, PLCs consist of a processor, input/output (I/O) modules, power supply, memory, and operating system. PLCs can be easily programmed.

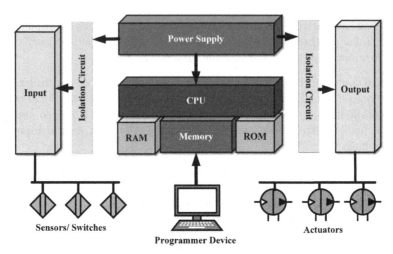

FIGURE 11.2: The Programmable Logic Controller (PLC) architecture and components.

Fig. 11.2 shows the various components of a PLC system. They can use both text-based as well as visual programming for network/device configuration. Text-based methods include instruction lists and structured texts. Visual methods include ladder diagrams, function block diagrams, and sequential function charts. PLCs use a cyclic scanning method to read statuses and data from its I/O modules and implement changes according to the implemented logic. PLCs can work with continuous as well as discrete signals.

11.3.1 Components

PLCs are composed of the following major components, which dictate its operations:

(a) Rack/Chassis: These act as the housing for the PLC components such as the power supply modules and the I/O modules.

(b) Central Processing Unit: PLCs have a microprocessor-based CPU. It reads data from the I/O modules and sends commands to them based on the implemented logic in it. It can have two types of processors – (1) a single bit processor (for logic functions) and (2) word processor (for processing text and numerical data).

(c) Memory: This works in conjunction with the CPU. Typically, two types of memory are present in a PLC – (1) ROM and (2) RAM. ROM is used for storing the operating system, application programs, and drivers. RAM is used for the storage of data and programming.

(d) I/O Modules: PLC I/O module acts as an interface between the CPU and physical devices. Devices such as switches, relays, motors, valves, and sensors can be integrated with this module to make them send/receive commands from the CPU. This module typically consists of an isolation circuitry, rectifier circuit, and a level detector circuit. The process devices operate at 220 VAC, which is converted to 5 VDC for logical operations at the CPU by this module.

(e) Communication Interface Module: These modules are crucial for enabling communication between remote PLCs and computers.

(f) Power Supply Module: The whole PLC setup derives power from a central location. This module enables the distribution of power throughout the PLC network. Typically, PLC operates on 24 V. This module also enables power conversion and AC-DC conversion for driving the processor and I/O modules.

11.4 SCADA

SCADA or Supervisory Control And Data Acquisition is a system of hardware and software components, which enable remote control, monitoring, and maintenance of local and remote industrial processes in real-time [171]. SCADA systems read and control installed devices such as sensors, actuators, pumps, and valves. The data is logged for historical records and visually summarized through an HMI interface. Networked SCADA systems rely on commonly available, vendor-agnostic communication technologies such as Ethernet. Present-day SCADA systems are robust enough to enhance efficiencies, reduce downtimes, and put up smart decisions based on input values from sensors. Fig. 11.3 shows a standard SCADA network and its components.

11.4.1 Components

The following are the basic components of a SCADA system:

(a) Programmable Logic Controllers (PLC) and Remote Terminal Units (RTU): PLCs and RTUs interface SCADA systems with sensors. The data from installed sensors, analog or digital, are read by PLCs and converted to digital signals. These are affordable, robust, and can be easily reconfigured. The control signals from SCADA are executed on the actuators through these units.

(b) Supervisory System: These units act as the supervisor or master of the SCADA network. They interface with the PLCs and issue control commands to them for execution at the actuators. These systems are configured for long-term and continuous monitoring of system errors and faults.

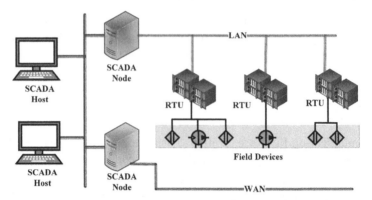

FIGURE 11.3: The SCADA architecture and components.

(c) Human Machine Interface: The HMI is used for enabling control and operations of the remote processes by human operators. The SCADA database linked to this interface provides visualizations, and linked software provides analytics and management capabilities to the human operators.

(d) Communication Infrastructure: The SCADA communication infrastructure is made up of a combination of wired and wireless connections. Typically, complex and massive SCADA systems rely on SONET and SDH for communication.

(e) SCADA Programming: The SCADA programming language uses either the C language or its derivatives. The programming helps in developing maps and diagrams for visualization and alert generation at the HMI.

Summary

This chapter provided an overview of some of the modern industrial control infrastructures. The basic overview of these systems has been covered, highlighting the crucial features and their formative components. These systems rely on various sensors, actuators, and data transmission mechanisms for their functioning. Depending on the inputs received from the sensors through the data transmission channels, the control systems generate decisions for maintaining the state of industrial operations through the actuators. Local as well as remote device installations, can be monitored, controlled, and maintained through these systems.

Exercises

1. Why are industrial control systems important?

2. What is a DCS?

3. How is a DCS different from PLC?

4. Differentiate between PLC and SCADA.

5. What are the differences between DCS and SCADA?

6. What is an HMI?

IIoT Analytics

the predominant forces acting upon it change, as shown in Figure 3. In particular, the effects of gravity (Fgrav) become negligible for radii of less than 0.1 mm, with surface tension (Ftens) and van der Waals (Fvdw) forces becoming dominant. In order to fully evaluate the challenges and potential solutions, it was first necessary to identify the individual assembly operations. To this end, the assembly operations proposed by Rampersad [7] were used to define the operations specific to this application. This was further enhanced by the use of the Design for Micro Assembly methodologies proposed by Ratchev and Koelemeijer [2]. As a result of this analysis, an assembly operation hierarchy was produced. This hierarchy is headed by the required outcome of the process, in this case, 'Micro-CMM probe assembly'. This outcome is broken down into two assembly tasks; each of these tasks is sub-divided into several assembly operations. Each of the assembly operations was analysed, with respect to the two core factors, as a means of determining the challenges faced and thus identifying a preliminary route for generating a solution. The results of this are shown in Table 1.

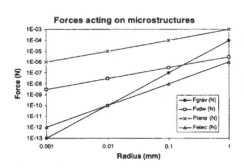

Fig. 3. The forces acting on an object with a radius of less than 1 mm

Fig. 4. The Zeiss NVision 40 FIB/SEM Crossbeam

2.2 Solution Development

With the challenges and requirements identified above and in previous trials [8], the next step in the work involved is identifying potential solutions to each operation. The first assembly task, hereafter referred to as Assembly Task 1, was to join the flexure and the shaft. The second assembly task, Assembly Task 2, is joining the sphere and the shaft. Both proposed solutions make use of the Zeiss NVision 40 FIB/SEM Crossbeam with integrated Kleindiek micro manipulators (Zeiss NVision), shown in Figure 4.

NPL provided the parts for all initial trials, namely: the piezoelectric flexure, the shaft and the probing sphere. However, the manufacturing process used by NPL could not guarantee the dimensions or even the presence of the hole in the flexure in which the shaft would be mounted. Therefore the preparations for Assembly Task 1, the joining of the shaft and flexure, were to test whether the Zeiss NVision could be used to machine the hole in the flexure.

Table 1. Summary of the key challenges and requirements for each of the assembly operations within the assembly tasks

Assembly Operation	Challenges	Requirements
1.1 Handle/Fix Flexure	• Very thin (15 μm) flexures • Delicate structure	• Keep electrical contacts clear • Minimise stress on flexure
1.2 Drill hole in Flexure	• Thin and delicate flexures • Comparatively large dimensions for FIB machining	• Do not damage flexure • Small tolerance for hole diameter
1.3 Handle/Fix Shaft	• Gravity no longer dominant force acting on part • WC is stiff but brittle	• Minimise acceleration and external forces • Preserve the delicate features
1.4 Insert Shaft into Flexure	• Delicate parts connecting at vulnerable point • Very restricted view	• As for 1.1 and 1.3
1.5 Join Shaft to Flexure	• Very low thermal capacity • Limited access to joint location	• Angle = 90° ± 0.29° • No distortion post-join • Joint resistant to fatigue
2.1 Handle/Fix Sphere	• Gravity no longer dominant force acting on part • Spheres unstable for gripping	• Sphericity of part is crucial • No localised damage or profile changes permissible
2.2 Handle/Fix Shaft	• Gravity no longer dominant force acting on part • WC is stiff but brittle	• Minimise acceleration and external forces • Preserve the delicate features
2.3 Insert Sphere onto Shaft	• Parts <1 mm: tension and van der Waals forces will affect alignment	• Near-perfect concentricity between sphere and shaft
2.4 Join Sphere to Shaft	• Very limited contact area • Low thermal capacity of glass prevents use of thermal processes	• No damage to parts • No joint degradation • No post-join deformation

For Assembly Task 2, it was proposed that the micro-manipulators within the Zeiss NVision should be used to deliver the probing sphere to the WC shaft for assembly. A plan was generated that would enable the existing tools to assemble the parts. Firstly, the shafts are loaded onto passive fixtures that retain them in the horizontal plane. The fixtures are then connected to the transfer shuttle for use in the FIB/SEM chamber. The sphere and SEM glue are loaded onto the same fixture for transfer into the chamber. The SEM glue is deposited into a small inverted cone machined into the surface of the fixture whilst the sphere is placed onto a small quantity of carbon. Once loaded into the chamber, the SEM beam is used for imaging to guide the operator during the assembly.

2.3 Existing Equipment

The Zeiss NVision is capable of performing several different tasks simultaneously upon samples held in its vacuum chamber. These include SEM, FIB, Gas Injection System (GIS), and manipulation via two Kleindiek nano resolution manipulators. The

tion to IIoT Analytics

chapter provides a preliminary understanding of the types
eed for analytics in industries.

readers will be able to get a flavor of the usefulness of an-
s, and the challenges in the implementation of analytics in
tries.

tionally, this chapter discusses the major stages for the de-
ent of analytics, machine learning, and deep learning in
tries.

applications of analytics in different industries are also out-
in this chapter.

oduction

istry 4.0 form a unique connected architecture, which has led
ncements in the industrial sector. The integration of opera-
rmation technologies results in the transformation of traditional
nart factories. The outcome of the fusion of various advanced
uch as M2M, complex analytics, AR and VR, and lean manu-
ms is a well-coordinated and connected system. The four main
f a smart factory are smart sensors, gateway, remote control
it, and communication unit [173]. The smart sensors connect-
ices and machines generate huge volumes of data in real-time
communication protocols. The gateways used for transmission
te network management and storage capabilities. The data col-
nsors are stored in the cloud or server, which can be retrieved
sory Control and Data Acquisition (SCADA) systems. Further,

:essing and analysis of the collected data provide meaningful insights,
.elps in making intelligent decisions. Without analysis, the usefulness
data can be compared to crude oil, which is useless until it is refined
in a range of highly usable and precious products. The analysis of
a helps to transform the data and extract relevant information from
eneral, data analysis is performed through the fusion of statistics/
iatics, coding, and business analysis skills. Different languages such
ion, R, and SQL are applied for data analysis. Data analysis is the
to analyze, inspect, transform, and remodel the data collected from
ory sites. Further, data analysis helps to predict faults and prescribe
ve measures, improve system efficiency, reduce downtime of machines,
iance the quality of products. This is achieved by giving priority to
s and financial decisions and analyzing the risks associated with them.
inufacturers can also modulate the speed and capacity of their assets
ive the predicted goals.

Necessity of analytics

inected industrial assets and devices sense and transmit data to the
: edge or cloud for further processing. Due to their connected nature,
cesses are monitored and can be directly operated by personnel, which
imotes the safety of workers in the plants. Further, the various health
caused to them, because of the risky operating conditions, can be
. The raw data is processed, stored, and analyzed to derive meaningful
and forecast future corrective actions. In addition to this, the associ-
processes with outliers and features can be developed. Various data
methods are used to analyze data, provide insights, and improve the
productivity of industries [172]. With the help of intelligent operations
timal decision-making, higher levels of automation can be achieved,
ers' demand can be modified, and new and improved products are
ctured. Therefore, the analysis of data helps to extract meaningful
tion from raw data, which identifies the hidden pattern in the data,
es the operational states of processes, predicts the performance of sys-
id prescribes corrective actions. For example, in case of any predicted
icy, a worker can remotely trigger alerts or shut down the process.
dicted data is applied for monitoring the health and safety of workers.
, the conclusion derived from the data in the form of statistics, trends,
terns helps in the execution of decision-making. This improves the op-
il efficiency of the entire plant. Further, cloud service providers provide
s of collected sensors' data as a service. Hence, predictive maintenance
trial assets can be offered as an additional service by service providers.

Analytics

ıt of the fourth industrial revolution or Industry 4.0 and IIoT,
sses and connected products, and smart devices have led to an
ı process efficiencies and created new business opportunities.
and actuators provide networked connectivity among the ma-
ices in the plant. Sensors, controls, and automation are present
es, but the integration of advanced technologies results in fur-
; the efficiencies of upgraded systems. The major challenge in
s the integration of data generated from heterogeneous sources
ıtly, its analysis. Data analysis helps to predict the downtime
ʰich is one of the important cost factors in industries. Small-
ːale industries do not have the necessary means to adapt ad-
ɔgies as it requires either replacing the existing infrastructure
hem. However, the adoption of long-term business models has
ıng by reducing risks and minimizing the probability of failure.
ics plays an important role in the improvement of overall ef-
h, and production in industries. The positive impacts of the
ıdata analytics in the industries are as follows:

ɔT-based applications in the industries generate a huge volume
ʰese data are analyzed in real-time using complex analytics-
ications.

generated from sensors may be used in a structured, unstruc-
ːemi-structured format. Data analytics help to analyze and ex-
mation from various data formats.

ytics help to predict the abnormality and faults in the system.
ɔation of analytics in the industries lead to flourishing businesses
ʋing product quality. Streaming, spatial, time-series, and pre-
ınalytics are used in various industrial sectors to recognize and
ıy malfunction and take corrective actions.

ɔegorization of analytics: IIoT and Industry 4.0
ıtext

łytics provide information about the operational and behavioral
ınected machines and devices and take accurate and optimal
ˌlytics used in industries can be categorized as:

ʳe analytics: In descriptive analytics, the previously stored or
collected data are used to describe the hidden pattern in the
˟ data streams may include the status and usage monitoring,
ıaly detection in various industrial processes.

dictive analytics: The expected outcome from the industrial processes
be identified by applying machine learning and statistical methods.
ne relevant uses of this method are the prediction of the amount of in-
tory utilized, customers' demand, system failure, and energy consumed
he industries.

scriptive analytics: The outcome obtained from the predictive analysis
sed in the prescriptive analysis for optimizing the industrial processes,
dict the downtime of machines, and avoid failures.

outcome of the analysis can be used for visual interpretation of pro-
hat enhance human understanding, improves the safety of individuals
ets, and facilitate process control. Further, there are certain types of
·s, which can be applied in the industries for improvement in the busi-
174]:

eaming analytics: This type of data analytics is applied for the analysis
uge real-time datasets. The analysis of these datasets provides infor-
tion for detection and evaluation of prioritized operations and take
nediate actions.

.tial analytics: In order to find the spatial relationship between the
ects and analyze the geographic relation among them, spatial analytics
sed. For example, the location of vehicles in a smart parking facility.

ne-series analytics: The relationship derived from the time-dependent
aset for exhibiting the hidden trend and pattern in the data set is
formed using time-series analytics.

Usefulness of IIoT analytics

ial analytics help to provide an improved operational platform, perfor-
and safe environment. IIoT analytics assist in the identification and
; of various information patterns under different conditions in the in-
. sectors. The entire step-by-step process of data analysis is termed
mic operation optimization. For example, the amount of electricity
ed in a power plant and demand of end-users fluctuates with seasons,
rice, amount of raw materials, and resources available. Further, based
forecasted information, the industrial processes and systems can be
ed.

patterns present in the collected dataset can provide meaningful infor-
Conventional analytics are used to predict and improve the perfor-
of businesses. However, the estimation of the probability of failure of a
e, asset, or industrial device, is a complex task. In industrial settings,
ntime of machines is an unplanned event. The prediction of the ma-
uptime, identification and prediction of faults, and minimization of the

FIGURE 12.1: IIoT application challenges.

or repairs falls under the domain commonly known as *condition-
ng*. Both cases of over- and under-utilization of resources may
ion of overall efficiencies of plants and increase the cost of pro-
wastes produced in the industries are minimized, and system
increased by using a lean manufacturing process. In addition
ling the maintenance of machines and devices, using *prognostic*
lepends on the information collected on the lifetime character-
s, detection of wear and tear, and fault conditions. Therefore,
ce of machines is performed within the estimated time, and
ects on production processes are avoided. Industrial *analytics*
to utilize the resources and workforce optimally, and further,
ficiency of machines. Predictive and prescriptive maintenance
a the analysis of raw collected data and extracts meaningful

allenges of analytics in industries

n of analytics in the industries modifies the operation of the pro-
ems. The heterogeneous type of smart sensor nodes deployed at
rial locations generates data in various formats, which may re-
s. However, there are certain necessary conditions to be fulfilled
and analysis of the data in industries. The various challenges
ion of analytics are illustrated in Fig. 12.1, which are discussed

ss: The outcome generated from the collected industrial data is
machines to update the processes automatically. Therefore, the
of data analysis is necessary such that undesired consequences
oided.

he time taken for performing analysis must be capable of accom-
the time for analysis of raw datasets and synchronizing them.
lly, the machines and connected devices should maintain relia-
quality of service in the various industrial processes within a
ime window.

ety: With the application of the analytics, the safety of the workers and rs can be improved. The interpretation and analysis of data generated) to protect the workers in various hazardous situations.

itextualize: The analysis of industrial data is performed in context to appropriate activity and observations in the plants. The proper understanding of the data and appropriate corrective actions help to derive meaning from the data generated from various industrial equipment processes.

tual-oriented: Industrial operations deal with domain-specific subjects, ch makes it essential to model the complex and hidden pattern in data appropriately. Further, to improve the accuracy of the outcome, ous statistical and machine learning operations are used.

tributed: Typically, the transformed smart industrial processes are distributed in nature, which necessitates the analytics applied on them to be ributed. Appropriate synchronization among the processes improves decision-making process.

eaming: The different industrial analytics run either as batch or continuous processes. Traditionally, an analysis of data is performed in batches improvement of the model. However, a huge volume of data is continually generated in the Industry 4.0 and IIoT scenario to provide real-time feedback to the industrial processes and improve them. Both continuous batch analysis models support these processes.

tomatic: As the industrial analytics applied to various processes and tems support continuous analysis of data, the outcome obtained should automatic, dynamic, and continuous.

mantics: The data collected from the industrial processes may be structured, unstructured, or semi-structured. The inference drawn from these a may be applied to increase the success and accuracy of industrial tems. The analysis of data in diverse forms is quite challenging.

Mapping of analytics with the IIRA architecture

ial Internet Reference Architecture (IIRA) is the common standard ture of IIoT, which provides new technologies, concepts, and applica-1 the context of functional components of the industrial system, IIRA ture is categorized into five domains – control, operations, information, tion, and business, as depicted in Fig. 12.2. Sensing, communication, uation are the functions of the control domain. Similarly, provisioning, ing of processes, diagnosis, and optimization, fall under the purview perations domain. The fusion, transformation, modeling, and analy-ata is done under the information domain. Further, the functions of

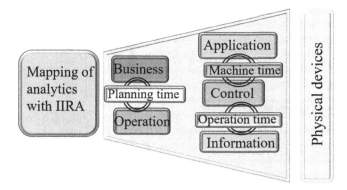

FIGURE 12.2: Mapping of IIoT analytics.

the application domain include logic, rules, and human-to-system interfaces. The business domain maintains and monitors the relationship with customers, assets, work planning, and scheduling.

Industrial analytics can be applied in the control domain present at the network edge, which takes about a few milliseconds or less to generate a response. Further, analytics can be applied in the operation and application domain to provide operational insights, facilitate predictive maintenance, and ensure optimal fleet operation. Analytics is applied in the business domain to predict the customers' demand, help in business planning, and provide meaningful insights regarding different business processes. Based on the demand query, either streaming or batch analytics are applied in the information domain [175].

The necessary conditions for applying analytics in industries are the availability and accessibility of real-time data collected from various industrial processes. Based on the type of analytics to be used, the data transmitted by sensors are stored, processed, and evaluated at the network edge or cloud. Further, data scientists may examine the data to compute correlations within them. After that, scientists use appropriate algorithms for clustering and classification of the datasets. Industrial analytics is capable of visualizing the collected data, connecting different applications, and processes within a common framework, managing information, and supervising the processes.

To determine where analytics are to be applied, certain factors need to be taken into account. The selection of the appropriate type of analytics to extract information from raw data is the first factor. Secondly, in the industries, response time acts as another crucial factor in determining the reliability and performance of the system. Thirdly, the fusion of data from one domain to another, and its analysis to derive valuable insights is also necessary. Further, as numerous heterogeneous sensor nodes are attached to the machines and devices, generating a considerable volume of data, proper network and infrastructure are required to support the combination of data from different

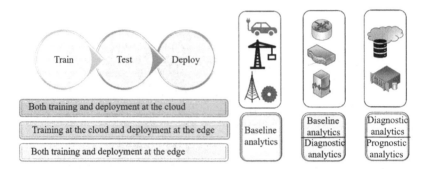

FIGURE 12.3: Stages of deployment of analytics.

domains. Finally, deriving the correlation among the different forms of data generated by sensors is a difficult task. The application of analytics close to the network edge may minimize the problems associated with such correlations.

12.2.5 Deployment of analytics

In industrial scenarios, most of the analytics in use are hybrid. Necessary analytics operations are performed at the edge of the network to reduce latencies [175]. The predictions, which are not time-sensitive, or the analysis of historical data, are performed in the cloud. As discussed in Fig. 12.3, there are three major stages to deploy analytics:

- To train a predictive analytics model.

- To test the trained model.

- To deploy the model to predict real-time data.

The training and deployment of an analytical model can be taken up in various ways. First, both training and deployment can be done in the cloud. For the training and deployment of analytics at the cloud, data generated from sensors at the network edge are transmitted to the cloud. Further, the outcome of the prediction is transferred to the network edge from the cloud. Similarly, in another approach, the dataset can be trained at the cloud, and the trained model can be deployed at the network edge. During the deployment of the trained model, both the cloud and the analytical algorithms need to be present at the edge. Lastly, for low latency applications, based on the computation capability of the edge nodes, both the training and deployment of the model can be done at the network edge.

12.2.6 Artificial intelligence

Artificial Intelligence (AI) is a domain of computer science, which deals with intelligent decision-making by machines and systems. The decision-making process of machines involves communication among machines with minimum or no human intervention [175]. The applications of AI in various industrial processes help to improve system efficiencies, upgrade product quality, and enhance accuracies. Machine learning and deep learning are the two important domains of AI. However, an ample amount of data must be accessible for performing analysis and predicting the state of the processes and systems.

(a) Machine learning: The training and deployment of models using machine learning are done through the collection of data, extraction of information from data using different algorithms, reducing the features, validation, and deployment of the model. Heterogeneous types of sensor nodes sense and transmit data to the network edge or cloud. The collected data is processed and analyzed to extract information and provide meaningful insights. Information is extracted from the data collected from multiple sources by applying feature extraction. Further, using this method, certain selected features are transformed into a lower dimension without affecting the original properties.

Depending upon the nature of information present in the real data, machine learning is categorized into two parts – supervised and unsupervised. In the supervised learning method, a labeled dataset is used as input. One of the essential highlights of the supervised learning method is the need for adequate data samples. Supervised learning methods incorporate training data from various sources to enhance data diversity for making the trained model robust and accurate. In contrast, unsupervised learning does not use labeled datasets as input. In order to detect the hidden pattern in the data set, unsupervised learning methods such as clustering of data – using various algorithms such as K-means and hierarchical clustering – are applied.

(b) Deep learning: Deep learning is performed through the collection of data, creation of the model, validation, and deployment of model. Motivated by the function and structure of a biological neural network, artificial neural network-based algorithms are popularly used for deep learning. These algorithms comprise of interconnected nodes arranged in the form of layers. The input layer is the first layer that interfaces with the input data. The last layer is the output layer, which provides the decision. In between the input and the output layer, there are multiple hidden layers, which assist in feature extraction. Deep learning requires an extensive dataset for training and is computationally intensive, typically requiring special processors known as GPUs. Based on the application type, two popular types of deep learning methods are – convolutional (CNN) and recurrent neural networks (RNN). CNN are mostly applied to the classification of objects or images. These networks are used for image recognition and classification. On the other hand, RNN is applied for the prediction of time-series information. The inputs and outputs are independent of time, or the method of arrival in the RNN. Further, RNN stores the previous state information and utilizes that information for making future state predictions.

12.2.7 Applications of analytics across value chain

With the integration of advanced analytics in industries, new value streams have been created, which help in improving product qualities, upgrading existing business policies, creating new business models, minimizing costs of process optimization, and enhancing process automation. The application of analytics in the industries include services from different components, such as data management and connectivity. The collection of raw data from various industrial processes, processing of data, and analyzing them, fall under data management. On the other hand, the storage, communication of data among different industrial processes, and relaying of decisions to machines/systems/processes fall under connectivity.

Analytics is widely applicable across various industries such as healthcare, retail, supply chain, manufacturing, and R&D companies. In R&D companies, product quality can be improved by incorporating the feedback received from the analysis of product characteristics. Similarly, in the manufacturing industries, predictive maintenance and decision support systems assist in reforming the industrial process parameters such as machine speed, product quality, and operation time of machines. Further, integration of analytics with logistics and supply chain results in the appropriate usage of resources, optimizes the inventory levels, and reduces transportation and fuel costs. The amount of revenue generated in the industries increases with the application of analytics.

Summary

1. Analysis of data aids in the extraction of hidden information from the collected raw data and prescription of appropriate corrective actions.

2. The application of data analysis helps to predict and prescribe faults, improves system efficiency, reduces downtime of machines, and upgrades the product quality in the industries.

3. Small-scale and medium scale industries do not possess the necessary facilities and skills to adapt advanced technologies. However,

the adoption of long-term business models reduce risks and the probability of failure in the machines.

4. The analytics applied for the analysis of data in industries can be classified as descriptive, predictive, and prescriptive. Further, streaming, spatial, and time-series analytics are also used to improve the business in industries.

5. The integration of analytics with industrial processes has certain challenges such as the correctness of data, safety, synchronization of data, and continuous analysis of data.

6. One of the essential conditions for the application of analytics in industries is the availability and accessibility of real-time data collected from industrial processes.

7. In the context of the industrial system, analytics are applicable in the control, operation, information, application, and business domain of the IIRA.

8. The three major stages for the deployment of analytics in industries are training, testing, and deployment of the model.

9. Machine learning and deep learning are the two important applications of AI in the industries, which help to improve the efficiency, product quality, and accuracy of the processes.

Exercises

1. What is the impact of data analytics in industries?

2. Give some industrial use cases of various analytics in the industrial sector?

3. How IIoT analytics help in various industries?

4. What is condition-based monitoring? How IIoT analytics assist in the same?

5. What are the various challenges faced in the application of analytics in industries?

6. Write the necessary conditions required for the application of analytics in industries.

7. How are the analytics mapped with IIRA architecture?

8. What is machine learning? What are the specific applications of ML in industries?

9. Write a case study on the application of analytics in healthcare.

10. What are the applications of AI in manufacturing industries?

13

Machine Learning and Data Science in Industries

Learning Outcomes

- This chapter provides insights into the basic concepts of machine learning and data science, and its applications across various industries.

- New readers will get an overview of various types of machine learning methods such as supervised, semi-supervised, unsupervised, reinforcement, and deep learning algorithms.

- Additionally, readers will get an overview of AI-based applications and the software platform/framework developed for storage, processing, and visualization.

13.1 Introduction

Artificial Intelligence (AI) is one of the vital components of IIoT and Industry 4.0. Outcomes derived from the data transmitted by heterogeneous intelligent machines and devices assist in the decision-making process in industries. The application of analytics helps to predict the faults and localize them in the machines and devices, improve the overall efficiency of the systems, upgrade the product quality, and improve the safety of machines and workers. Typically, analytics used in the industrial applications are of three types – predictive, prescriptive, and descriptive. However, streaming, spatial, and time-series based analytics are also applied to industrial data. In the era of the fourth industrial revolution, AI has proved to be a powerful tool in the reformation of the industrial processes[178].

Data science is a scientific approach to provide meaningful insight from data using algorithms, methods, and processes. In order to process, analyze, and store a huge volume of data generated from industrial processes, data science is indispensable. The raw data collected from various industrial processes, machines, and devices may be in structured, semi-structured, and unstructured form. The information extracted from the data collected helps AI to identify hidden patterns and derive meaningful insights from them. AI is the capability of machines to operate and act intelligently, bordering toward human intelligence [184]. Machine learning (ML) and deep learning are the two common tools of AI that assist in making industrial technologies smarter. ML is a subset of AI, which predicts or provides decisions based on collected data. Similarly, deep learning is a form of ML based on the concept of dense artificial neural networks (ANNs). The relationship between AI, ML, and deep learning is depicted in Fig. 13.1. Typically, AI can be categorized as strong and weak AI. Strong AI is designed to be context-aware, cognitive, and capable of making decisions on their own. In contrast, weak AI is mostly dependent on algorithms and programmatic responses. For example, Alexa is a virtual assistant, developed by Amazon for voice interactions, setting alarms, music playback, and providing real-time information. However, Alexa is based on weak AI and processes data using a request-response behavior.

13.2   Machine Learning

ML is the science of algorithms and statistical models applied for the execution of a specific task. According to researchers, ML is the best method to develop human-level AI. Arthur Samuel coined the term ML in the year 1959. Later, Tom. M. Mitchell defined machine learning as, "A computer program is said to learn from experience $\mathbf{E}$ for some class of tasks $\mathbf{T}$ and performance measure $\mathbf{P}$ if its performance at tasks in $\mathbf{T}$, as measured by $\mathbf{P}$, improves with experience $\mathbf{E}$." Further, Alan Turing proposed the concept of intelligent machines and their properties in his paper 'Computing Machinery and Intelligence' [176]. In his paper, Turing asked the question, "Can machines think?" As this question seemed illogical and vague, it was reframed as "Can machines do what we can do?" In his paper, he discusses the different characteristics which a thinking machine should possess.

In the IIoT and Industry 4.0 environment, a considerable volume of data is generated in structured/unstructured format. The complete understanding of the type of fields or attributes present in the data is necessary for preprocessing the data. For example, the data object in a hospital database may be patients, doctors, or staff. Therefore, the patient's ID, age, weight, and disease are its attributes. Typically, there are four types of attributes present in data – nominal, binary, ordinal, and numeric. Nominal, binary, and ordinal attributes are qualitative in nature. Only the numeric attribute is quantitative.

Therefore, as the numeric attributes are measurable, they are represented in the form of an interval or ratio scale. The interval-scaled attributes are measured in the scale of equal points, while the ratio-scaled attributes are numeric attributes with a zero point. A nominal attribute represents the value of some category, code, or state of data, and hence, are also known as categorical attributes. Further, binary attributes are represented either as 0 or 1 (e.g., in the case of an HIV test, the patients' results can be represented either as 'Positive' or 'Negative'). The ordinal attributes are represented in the form of meaningful rank or order. Further, the data are also represented as discrete or continuous attributes. Before preprocessing data, it may be graphically represented in the form of scatter plots, box plots, and histograms. Based on the type of data, various techniques such as pixel-orientation, geometry-based, and icon-based techniques are used for visualizing it.

In order to apply ML to predict and prescribe decisions, a significant amount of data is collected at the beginning. Thereafter, the collected data is prepared for analysis. In industries, as heterogeneous types of data are commonly encountered, specific methods such as standardization and normalization are used to prepare the dataset prior to analysis. Typically, the collected data is split into a 70:30 ratio for obtaining the training and testing data. Based on the sample training dataset, a mathematical model is established. This mathematical model is known as *training data*, which is responsible for predicting or making decisions. After that, the evaluated model is tested with the testing data. Subsequently, the predictive accuracy of the developed model is checked, and the training error is computed to check whether the model is overfitting or underfitting. During the training phase, if the model adopts a certain pattern from data that is not characteristic to the data set, overfitting occurs. Typically, overfitting is witnessed in complex models. On the other hand, for a simple model, underfitting may occur, which leads to poor prediction. The application of ML in a domain/task helps to extract complex and hidden patterns from the task-specific data, which are then utilized to come up with mathematical models that can make intelligent decisions for those tasks.

Expectation-maximization algorithms are applied for real-world ML applications such as density estimation and clustering for missed values. In order to predict the values, the maximum likelihood/maximum posteriori values of a statistical model are computed iteratively. Typically, in most of the cases, these statistical models depend on latent variables.* This algorithm computes the decision probability in two steps – in the first step, the values of variables are computed, and in the second, optimization is performed until the convergence criterion is met. Further, the first step is termed as the expectation step

---

*The values of the latent variables do not depend upon observed variables. Observed values of different variables influence latent variables. For example, the health criticality of a patient in the IoT scenario can be predicted from the values of the sensors attached to the body of a patient

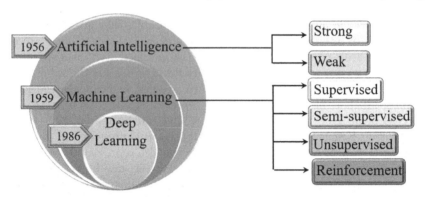

FIGURE 13.1: AI, ML, and deep learning.

(E-step), and the second step is known as the convergence step (C-step) [191]. The E-step and C-step are easily implementable. However, the convergence occurs slowly in this algorithm.

13.3 Categorization of ML

ML is essential for decision-making and requires many resources. Various types of ML algorithms are represented in Fig. 13.1. Depending on the algorithms used to establish the mathematical model, ML is categorized as follows:

- Supervised learning: This type of learning utilizes a sampled/labeled data set. Supervised learning algorithms require the direct supervision of the dataset and affix certain limitations. The incoming dataset is labeled based on past samples. Then, the algorithm recognizes and classifies these labeled datasets. Fig. 13.2 illustrates the various examples of supervised learning. The two major processes of supervised learning are classification and regression. Classification is the form of analysis of data that extracts meaningful information from the data and segregates them into classes based on similarities. Learning/training of the data is the first step for data classification. Thereafter, the class labels are predicted from the dataset. The segregation of nominal, binary, and ordinal data attributes utilizes classifiers. In contrast, regression analysis is applicable for numeric attributes. The most popularly used supervised learning algorithms are – linear regression, logistic regression, random forest, support vector machines, neural network, and decision trees. Supervised learning is used to predict prices, forecast sales, commerce, and stock trading.

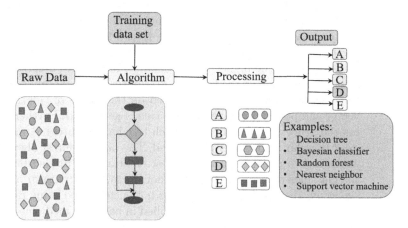

FIGURE 13.2: Supervised learning.

a. Linear Regression: Linear regression is a form of ML applied for the selection of target values. The values of the dependent variable are determined depending on a given independent variable. It is mostly applied to forecast the values of variables and detect the relationship between the variables. As the regression technique detects a linear relationship between a dependent variable (output/response) and set of independent variables (input/predictor variables) based on a discovered holistic trend, therefore, the function is termed as linear regression. Mathematically, linear regression is represented as

$$y = \beta_0 + \beta_1 x_1 + \beta_2 x_2 + \cdots + \beta_n x_n \qquad (13.1)$$

where $\beta_1, \beta_2, \cdots, \beta_n$ are the coefficients of the inputs $(x_1, x_2, \cdots, x_n)$ and y represents the response variable/output.

b. Logistic Regression: Logistic regression is applicable when the dependent variable is binary/dichotomous. This form of regression analysis illustrates the relationship between a dependent variable and one/multiple nominal, binary, ordinal, and numeric independent variable/s. The major difference between logistic and linear regression is that dependent variables are binary in the case of the former, while the dependent variables are continuous, in case of the latter. Additionally, in the case of linear regression, a linear relationship is maintained between the dependent and independent variables, which is not necessary in case of logistic regression. The logistic function is also known as the sigmoid function. It is described as an S-shaped curve.

c. Random Forest: Random forest is an ensemble-based method of ML algorithms applied for classification and regression. The fundamental concept of random forest is that each of the classifiers in the ensemble is assumed to be a decision tree, which combines to form the forest [187]. Moreover, the higher the number of trees in the random forest, the higher is the accuracy. In order to ascertain the split at each node of the tree, the attributes are randomly selected to generate the decision trees. Based on the values of the random vector sampled independently, the tree is generated. Further, during the process of classification, each tree has an outcome, and the most popular among them is selected. The random forest algorithm works in two phases – random forest creation and prediction from the classifier developed [189]. Further, in case of any form of classification, the overfitting problem is not encountered using random forests. This classifier is mostly applied in banking, medicine, stock market, and e-commerce applications.

d. Support Vector Machine: Support Vector Machine (SVM) is a method used for the classification of linear and non-linear forms of data. SVM can be used for both classification and regression. Classification is performed by SVM using hyperplanes. The hyperplanes maximize the margin between two classes. The vectors which define a hyperplane in a two-dimensional space are termed as support vectors. In addition to this, hyperplanes also map the data into higher dimensions, where the linear decision space can be easily divided and classified. In the case of linearly separable optimization problems, the generalized form of the classifier function, $f(x)$ is represented as:

$$f(x) = w^T X + b \qquad (13.2)$$

where w and b are the weight vector and the bias such that $\forall x_i \in X$. However, in case of non-linear data, slack variables are introduced. The generalized classifier function $f'(x)$ for non-linear variables is:

$$f'(x) = f(x) + w' \sum_{i=1}^{m} \eta_i \qquad (13.3)$$

where η_i denotes the ith slack variable introduced in the function and w' represents the relative weight factor. With the introduction of slack variables, some of the instances may fall off the margin. The points which are present within the margin of the hyperplane are subjected to a penalty [188]. If any of the points are outside the margin of the hyperplane, no loss of generality occurs.

e. Decision tree: Decision tree is a form of supervised learning used for both classification and regression. A decision tree is represented in the form of a tree-like structure where the initiating/topmost node is known as the root node. The internal nodes of the tree represent the attributes, and the leaf nodes denote a class label. Each branch of the decision tree represents an outcome of the classification procedure. Further, in a decision tree, the path is traced, starting from the root node and ending at the leaf node. This process helps in the prediction of classes. Moreover, decision trees can be applied to multidimensional data. The steps of learning and classification in decision trees are simple compared to the other supervised learning algorithms. CART is one of the methods which apply the decision tree algorithm for attribute selection. Further, to measure the impurity in the dataset or training samples, the Gini index is used.

f. K-Nearest Neighbor (KNN): The KNN algorithm is a simple ML algorithm applied for solving classification and regression problems. The datasets are classified based on a measure of similarity. The 'K' in KNN represents the nearest number of neighbors of the labeled dataset, which are checked against the testing data. The distance between the trained instances and input labeled dataset is usually computed using *Euclidean distance*. However, Manhattan and Hamming distances may also be used to compute the distance between the training and test datasets. Let us consider (x_i, y_i) and (x_j, y_j) represent the points of the training and test dataset. Therefore, the distance (D_{ij}) between these set of instances is mathematically represented as:

$$D_{ij} = \sqrt{(x_i - y_i)^2 + (x_j - y_j)^2} \qquad (13.4)$$

Based on the value of K, the nearest neighbors of the test data are observed. Finally, the test data is compared to the most similar class to which it belongs. Therefore, the selection of the value of K is a complex task. There are several industrial applications of KNN. For example, when product A is purchased, other products that are likely to be bought/related with product A are recommended on the e-commerce websites, to tempt/attract customers to buy the combo, and increase the average amount for the products ordered. Similarly, as a huge volume of data on diverse topics exists on the Internet, which can be searched using KNN using rudimentary keywords [190].

- Semi-supervised learning: Semi-supervised learning algorithms are positioned in the middle of supervised and unsupervised algorithms. The semi-supervised algorithms use a limited set of labeled data to train their models. Fig. 13.3 depicts the schematic representation of the semi-supervised learning. A partially trained model can label/transform the unlabelled data to pseudo-labeled data. Labeled and pseudo-labeled data are collectively used to identify assets and classify data. Semi-

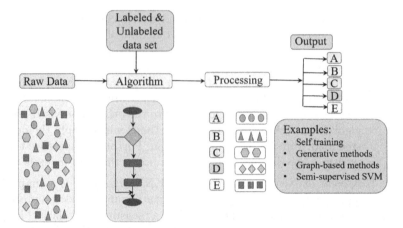

FIGURE 13.3: Semi-supervised learning.

supervised learning algorithms find widespread use in legal, healthcare industries, and speech analysis.

- Unsupervised learning: Unsupervised learning uses unlabeled datasets and is not directly controlled by the developers. The outcome of the algorithm developed is quite unpredictable. Unsupervised learning is applied to find information structures, extract valuable information, and detect hidden patterns in data. The application of unsupervised learning improves the efficiency of the processes. Clustering of data and reduction of data dimensions are the two most commonly used methods of unsupervised learning. The clustering method is used to group the data of a similar pattern. K-means clustering, Principal Component Analysis (PCA), and t-Distributed Stochastic Neighbor Embedding are the most popularly used unsupervised learning methods, as shown in Fig. 13.4. Digital advertisement technologies use unsupervised learning to predict customer behavior.

  a. K-means: K-means clustering is a form of unsupervised ML method used to cluster samples, specifically numerical attributes. The 'K' in K-means denote the number of clusters/groups which are to be identified from the data. Initially, the clusters are randomly selected, and coordinates of the centroid of clusters are identified. These randomly selected centroids are the initial points required to form a cluster. Thereafter, the distances between the different coordinates in the clusters are usually computed using squared Euclidean distances. The minimum distances are selected, and clusters are formed again. The clustering process discontinues when the centroids remain unchanged, signifying certain points lie within the

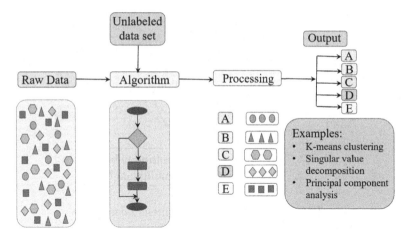

FIGURE 13.4: Unsupervised Learning.

cluster, and maximum iterations have been performed. One of the major advantages of the application of K-means is that it can be easily implemented to analyze large volumes of data and extract meaningful information from them. K-means clustering is popularly applied in various industrial applications such as the separation of customers based on the type of items bought by them, frequency of their visits to specific websites, and the number of transactions performed by them [192]. Additionally, K-means clustering is also applied as a featured learning step in case of semi-supervised or unsupervised learning.

b. K-medoid: K-medoid is a clustering algorithm used to minimize the distance between some randomly selected data points in a cluster. Medoids are points inside a cluster, where the distance with other points in the cluster remains minimum. Both K-means and K-medoid are clustering algorithms, which aim to minimize the distances between labeled points in a cluster. K-medoid selects the data points as centers and randomly computes the distances. However, in the case of K-means, the center of a cluster is not always one of the data points. The K-medoid clustering algorithm operates in two steps – *build* and *swap*. In the build step, the 'k' center points, known as medoids, are sequentially selected. Various distance measurement methods such as Euclidean, Manhattan, and Hamming distances are applied to measure the distance relative to other associated points. In the swap step, the objective function is minimized by interchanging a selected object with another unselected object. K-medoid helps to minimize the problem of outliers through the adjustment of the points inside a cluster.

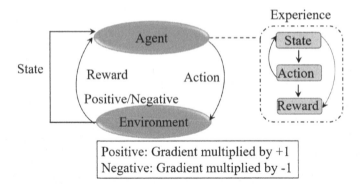

FIGURE 13.5: Reinforcement Learning.

- Reinforcement learning: Reinforcement learning helps to develop a self-sustaining system through a continuous sequence of trials and errors, depending on the combination of labeled data. In the case of neural networks, the error gradient can be computed using the backpropagation algorithm on the output of a stage. However, in supervised learning algorithms, a labeled dataset is necessary to train the model. In contrast, in reinforcement learning, an agent learns to develop the outcome itself. The input frames are transformed into output actions through a set of policies. The primary motivation of reinforcement learning is to optimize the policies and receive as much reward as possible. After the execution of a specific task, reward signals are provided, as illustrated in Fig. 13.5. These reward signals act as a navigation tool for these algorithms. The reward signals can be positive or negative. A positive reward signal motivates the agent to continue with the algorithm. On the other hand, a negative reward signal prohibits the agent from performing certain actions and recommends corrective actions. During the training phase, the positive reward signals are more likely to occur, and negative signals are filtered out. For example, the development of the *Grasp and Stack* process of a robotic arm utilizes reinforcement learning. Further, reinforcement learning is also applicable to various industrial automation systems.

13.4   Applications of ML in Industries

In industrial organizations, ML is used to predict and localize faults, improve efficiencies, and upgrade the quality of products. Further, the outcome of analytics assists in improving the security of systems and promotes automation of processes. Some popular applications of machine learning ML across various industries are as follows:

(a.) C3IoT: C3IoT provides cloud platform-as-a-service to the end-users. The end-users of C3IoT are various industrial sectors such as manufacturing, oil & gas, retail, transportation, banking, and healthcare. C3IoT provides different types of AI-based applications, which comprise a model-driven software and architecture. The software platform assists developers to access the capabilities, storage, processing, visualization, data exploration, and ML. The C3IoT architecture transfers complete applications and microservices. These applications include data integration, management, and processing capabilities. The AI-based C3IoT architecture comprises data abstraction, processing, and security layer. The complete C3IoT framework speeds up the development of applications and delivers those applications to industries [179].

(b.) GE Predix: Predix analytics platform helps to provide specific business outcomes. The Predix platform puts forward a unique framework to create ML analytics. Further, this platform helps to predict failures, automate processes, and improve overall efficiencies. When these analytics are applied to real-time data, the outcome helps to identify anomalies or faults in the system. The outcome also predicts any upcoming critical event and predictively generates an alert if maintenance is required. The analytics platform of GE is used across different industries such as food and beverage, wastewater treatment, equipment manufacturing, chemicals, automotive, steel, and paper manufacturing industries. The Predix platform is easy to deploy at the edge as well as at the cloud. The industrial-grade analytics and framework assist in the creation of reliable digital twin models [183].

(c.) Oracle: The analytics cloud service provided by Oracle is unique. The Oracle cloud analytics platform gives end-users a complete, connected, and collaborative platform. This cloud platform is applicable to various industrial processes and decision-making at the cloud and data centers. Oracle analytics platform provides various functions such as self-service visualization, advanced analytics, inline data preparation, and self-learning analytics. The analytics help to provide proactive insights. The customers have the flexibility to select the pay-as-you-go model, or monthly flex model, as per their requirement. The payment plans vary based on the type of product selected. The pay-as-you-go model allows customers to use the Oracle cloud analytics platform on an hourly payment basis. On the other hand, the monthly flex model provides flexibility to the end-users to make payments on a monthly basis [177].

13.5 Data Science in Industries

In the IIoT and Industry 4.0 scenario, a prodigious amount of data is generated from connected machines and devices. The real-time data generated helps to optimize processes, predict the downtime of machines, improve the quality of products, reduce wastes, and minimize production costs. Data science is the method of focusing on data and applying mathematical tools to find the hidden pattern in the data. In real-world scenarios, such as General Electric, experts apply real-time information at a common point of physical modeling of the system, digitize the processes, and apply traditional techniques [186]. Further, the application of data science in the transportation sector of GE leads to savings of about 8–10 million dollars, annually. Fig. 13.6 shows the various applications of data science and their benefits in different sectors. For example, the application of data mining and ML helps in formulating business strategies, and the application of AI and neural networks assist in the management of stakeholders. Further, data visualization enables visualization of hidden patterns in data. Data modeling and hypothesis testing assists in the communication of business and analysis of the product developed. The complex data analytics and big data assists in exploring new opportunities. The application of prescriptive and exploratory analytics on the collected data provides meaningful outcomes.

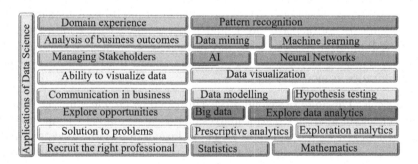

FIGURE 13.6: Applications of data science.

The various advantages of the application of data science in industries are as follows:

(a) The collected data is processed, and meaningful insights are derived. Further, the insights are transformed into actions to achieve appropriate business outcomes.

(b) The system throughput improves, quality of product enhances, and other general problems related to the manufacturing sites can be avoided.

(c) Data science helps to fabricate solutions for data-driven management and predictive analytics.

(d) Through proper optimization scheduling, field service operations can be improved.

(e) The probability of failure can be predicted from the interpretation of data patterns.

(f) With the help of prescriptive models, corrective measures can be identified.

(g) Complex mathematical algorithms are derived from empirical knowledge of data and are made functional through IIoT platforms.

The application of data science in industries provides various outcomes such as upgrading businesses through analysis of collected data, predicting consumed power before providing the solution, and accelerating businesses through data science solutions. Additionally, the outcomes of analytics can be used to train executives and project business outcomes. The exploration of the outcome of data science in an organization may span days to weeks. This process helps to prioritize the KPIs of the organization, which provides an optimized insight from the collected data. Further, based on the data, these KPIs help to develop a feasible and robust solution. The basic version of the data science survey may take one week to a month, at most. During this process, a team reviews the selected data sources, evaluates the collected data, performs primary tests to check the data quality and interaction among variables. After that, data science experts develop data-based solutions, which take almost 12 weeks. The investigation of data involves assessing the quality of data, finding a relationship among the variables, and analyzing their relationships with the KPIs. Further, the experts verify the developed model. Based on the outcome, the executives are given training, which takes about 2 days. These trainings are designed to impart the knowledge of the basics of ML, regression, and univariate analysis among the executives. With the help of data science, on-site development and improvement of algorithms are possible. Therefore, the data collected from different industrial sectors are appropriately utilized to create new business values.

13.6   Deep Learning

A deep learning algorithm is a specific type of ML that is primarily based on ANNs. Typically, ANNs are represented as an interconnected group of nodes, which draw inspiration from the structure and functioning of the human brain. These interconnected nodes in an ANN are known as artificial neurons. Mathematically and functionally, these artificial neurons are similar to the neurons in a biological brain. Fig. 13.7 shows an ANN with the input layer, hidden layers, and the output layer. Synaptic weights are provided to interconnect the input, hidden, and output layers. Each of the neurons in ANN is assigned a weighted value. The weights are updated between the input and hidden layer, and the hidden and output layer. The process of weight updating is done using backpropagation, or radial basis method. In the backpropagation method, the gradient descent method is applied to update the weights. Based on the rate of error computed in each of the epochs/iterations, these weights are updated. However, only the weights between the input and the hidden layer, and biases are adjusted. The appropriate weight update helps to reduce the rate of error. Further, a neuron with higher weight possesses different levels of influence on the preceding layers. These high-valued weights are multiplied with the larger activation functions. The last hidden layer or final layer combines the weighted inputs to produce the final output. The same process repeats recursively to adjust the weights and biases. However, the entire process of weight adjustment is time-consuming. The training data is randomly shuffled, divided into small batches, before updating the weights to fasten the update process. This approach improves the computational speeds significantly.

In the biological context, synapses are structures that help to transmit the signals from one neuron to another in the brain. Similarly, in ANN, each neuron communicates with other neurons by applying signals. Further, due to the ANN-based structures, deep learning can handle a high dimension of data. In the traditional neural network, the number of hidden layers may vary from 2 to 3 [185]. However, in the case of deep learning, the network may have up to 150 (or more) hidden layers. Unlike ML, deep learning does not need algorithms to specify the steps of data processing and analysis. The performance of deep learning algorithms improves with the increase in the volume of data. Further, in ML and AI, there is a presence of potential bias in the dataset and algorithms. It is necessary to mention the features of the data in ML explicitly. Therefore, to overcome such problems, a deep learning method is best suited for computation and analysis of an enormous volume of data.

The deep feedforward neural network is the simplest form of neural network. These forms of neural networks are also known as a multi-layered neural network. As the name indicates, the information in a feedforward neural network moves in the forward direction from the input layer to the output layer, through the hidden layers. When a feedback loop is added to the feedforward neural network, the neural network is known as a Recurrent Neural Network (RNN). In an RNN, the interconnection between the nodes forms a directed graph. A fixed time-series data is provided as input, which gives a fixed size output in a feedforward neural network. Further, the memory size/storage is another constraint in the feedforward neural network. RNN overrules these limitations of the fixed-size data and memory. The present state of input in RNN is a recursive function of the previous state and input at the previous time instant. In multilayer RNN, the output serves as the input to the new layers. Typically, in neural networks, the weights are updated, and the model is trained to apply backpropagation. The RNN adjust the weights and train the model using backpropagation through time. Therefore, the loss is computed using the output, and the weights are updated in the previous state.

Another form of the neural network is the Convolutional Neural Network (CNN), which is applicable mostly for image analysis. CNN is a specialized form of ANN, capable of detecting patterns and extracting hidden information from data. The main feature of CNN which distinguishes it from other neural network models is the presence of hidden layers known as convolutional layers. The basis of CNN are these convolutional layers, that receive inputs and transform them to the next layer after performing a convolutional operation. In the convolutional layers of CNN, the inputs and outputs are masked by the activation function and backpropagation is applied for final convolution to accurately update the weights. For example, there may exist multiple edges, textures, shapes, and objects in an image. Various types of filters are used to detect these 'patterns.' The deeper the network, the more finer is the analysis of the images.

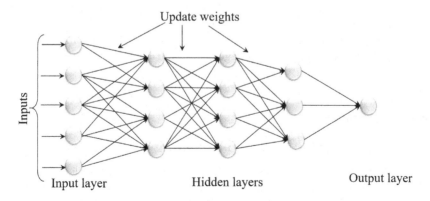

FIGURE 13.7: Artificial Neural Network (ANN).

The learning methods used may be supervised, semi-supervised, or unsupervised. To achieve the desired outcome, multiple layers are used in deep learning, as in the ANN network. These multiple layers help to extract higher-level features from the input. For example, in the case of image classification, deep learning algorithms require few thousands of images for acceptable accuracy. Training of a deep learning model requires a huge volume of labeled data and robust ANN architectures, which extract features directly from the data. Therefore, in the absence of labeled data and high-performance Graphics Processing Units (GPU), regular ML algorithms are applied to the data instead of deep learning approaches. A high-capacity GPU reduces computation and processing times in deep learning. Deep learning methods use an end-to-end learning approach hierarchically. The complex tasks are separated into simplified tasks to process these tasks rapidly. In order to perform object classification, three common methods are used in deep learning, which are as follows:

(a) Training from scratch: To train deep neural networks, a large labeled dataset is provided as input. After that, the network architecture is designed, which provides the features and model. This object classification method is suitable for a large number of output categories. Moreover, these networks may require days or weeks for training.

(b) Transfer learning: This process of training a deep learning model requires fine-tuning a trained model. Practically, most of the deep learning applications utilize the transfer learning approach. This approach requires less amount of data. Therefore, the time required for processing and computation may reduce from hours to minutes, and from weeks to days.

(c) Feature extraction: This is a specialized model of ANN. The layers comprise certain features extracted from the images which are provided as input. These features can be analyzed at any time instant during their training. Further, these features can be provided as input to any ML model, such as SVMs, linear regression, and logistic regression.

13.7   Application of Deep Learning in Industries

The industries which provide preliminary services such as electrical power transmission, airports, dams, and natural gas delivery form the utility sector. Regular inspection and maintenance of the utility sector are crucial for uninterrupted operations. For example, any fault in the overhead power transmission lines may result in forest fires, power outage, and may result in blackouts over large regions. However, analysis of collected data and prediction of errors may result in improvement in the overall efficiency of these industries.

The applications of deep learning in some industrial sectors are discussed as follows:

(a) Toshiba: Around 2 billion data collected from various connected devices are analyzed by skilled professionals for visualization, improving productivity, and enhancing product quality. However, with the integration of Industry 4.0-based advanced technologies, the amount of data generated has increased due to the increase in production volume and process complexity. The data generated vary in terms of type and format. To extract the pattern from data and predict the solution, deep learning is applied. As per a study related to domestic AI, the amount of data generated will reach by 87 trillion yen by the year 2030, which is 23 times the data found in the market for analysis in 2015. Further, the organization is searching for new automation techniques and improved solutions. Deep learning is applied to improve productivity through the analysis of data collected from wearable devices. Similarly, the data of the movement of the worker's arm is accumulated and analyzed to estimate workers' behavior. The efficient generation of power is done with precisely forecasted values of demand, weather conditions, and faults at multiple points with AI applications. Grid Data Lake is one of the products of Toshiba, which provides a scalable architecture for storing, processing, and analysis of data [182].

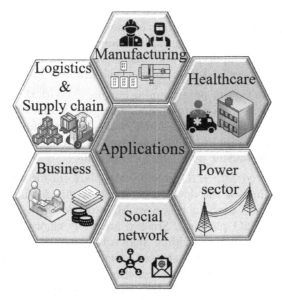

FIGURE 13.8: Applications of saltys.

Saltys is the AI analysis services provided by Toshiba to accomplish accurate identification, prediction, factor estimation, failure sign identification, and evaluation of device behavior. The key features of Saltys are the analysis of the big data, attaining highly accurate results with trained models, and intuitively explaining the abnormal factors. Saltys is widely useful across different industries such as the power sector, logistics and supply chain, healthcare, and manufacturing plants, as shown in Fig. 13.8. Saltyskata is a type of AI analysis service delivered by Saltys to the customers in the form of Software-as-a-Service (SaaS). This platform aids in inventory optimization, prediction of failures, and creates a failure model from the past data collected. Saltyskata prediction platform helps to truncate the costs associated with storing and disposal of excessive maintenance parts and reduces the amount of inventory required in the processes [181].

(b) H2O: The H2O is an open-source, distributed, and scalable ML platform. This platform provides various products such as sparkling water, H2O open source optimized for NVIDIA, and H2O ML platform. H2O integrates with the open-source platform, Spark, for its product sparkling water. Further, this organization is one of the leading AI technologies and product providers to other industries. Almost more than 18,000 companies around the world utilize the H2O platform. This platform is also very popular among the industrial and research groups using R and Python languages [180]. The various features of H2O are as follows:

- Algorithms: The algorithms are developed from the primary level for distributed computing. Additionally, these approaches are also applicable for supervised and unsupervised approaches such as random forest, generalized linear model, and gradient methods.

- Access to various programming languages: With the help of R, Python, and other common programming languages, models are prepared using the H2O platform. Further, the H2O platform is also used to develop a graphical notebook-based interactive user interface, which is not dependent upon any form of code.

- Automated ML: H2O provides the platform for automatic training and tuning of the developed models within a time duration specified by the user. In this way, ML workflow can be automated using the H2O platform. Stacked ensemble algorithms automatically train a group of individual models to generate the predicted ensemble model. The H2O's stacked ensemble algorithms are based on the process of an optimal combination of prediction algorithms known as stacking. Further, stacked regression utilizes the individual tasks used for cross-validation of data.

- Distributed processing: In order to process data faster, the H2O platform organizes data serially between nodes and clusters. Distributed

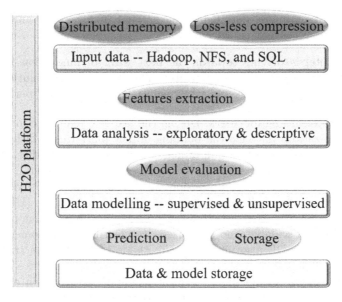

FIGURE 13.9: H2O platform.

processing of big data helps to deliver data faster up to about 100 times. Further, this enables in optimizing the operational efficiency without deterioration of the computational accuracies involved.

- Simple deployment: The H2O platform can be easily deployed for provisioning fast and accurate services.

The H2O platform provides the infrastructure for processing and computation of big data. The schematic diagram of the working principle of the H2O platform is shown in Fig. 13.9. Due to its distributed nature, the H2O architecture is capable of taking data as input from various sources and file systems such as HDFS, NFS, and SQL. After that, exploratory and descriptive analysis of the input data is performed. The designed model is evaluated and analyzed using supervised and unsupervised algorithms. Finally, the outcome is predicted from the stored data. The H2O platform is widely useful across different industrial sectors such as healthcare, insurance, marketing, telecommunication, retail, and manufacturing. This platform helps to improve the customers' experiences, minimize wastes, assist in improving business, and keeps pace with the modern challenges. For example, the various use cases of healthcare are the prediction of the transfer of a patient from ICU, upgrading the clinical workflow, and prediction of infections related to a hospital. The application scenario of the H2O platform in the field of manufacturing is predicting the manufacturing design, predictive maintenance, and optimization of the supply chains. In retail industries, the H2O platform is widely used for the recommendation of products, optimizing the amount of inventory, and providing products at optimal prices.

(c) Intel: Intel is driving towards the next wave of innovation for the development of an AI-based platform. The Intel selected solutions aim to eradicate presumptions and verify solutions for real-world performance.

(1) Deep learning for Spark: During the last few years, the industrial organizations have led the building of new platforms, which allow processing and computation of huge volume of data. Further, the analysis of data using AI helps to improve businesses. To standardize big data storage and computation capabilities, Apache Spark assists in solving the challenges associated with deep learning (DL) and big data. BigDL is a distributed deep learning library that provides an efficient, scalable, and optimized deep learning development environment. BigDL also allows new methods of DL for the training and development of clusters. BigDL is supported by Analytics Zoo, which provisions a unified AI platform pipelined with certain use-cases to simplify the development of AI solutions. The solution selected by Intel for BigDL helps to optimize the price and performance while minimizing the evaluation time. The comprehensive solution provided by BigDL offers the following:

(i) Storing and processing DL algorithms for the future with scalable infrastructure and computation power.

(ii) Providing infrastructure with multi-purpose hardware for the management of a verified and tested solution to simplify deployment.

(iii) Accelerating the time required to market solutions that are optimized for the development of essential products.

(iv) Providing the platform for performing analytics on the stored data.

(2) AI inferencing: The operational necessities of any organization directs them for the periodic upgradation of hardware. However, the investment in hardware, which is not multi-functional, may not fulfill the requirement of the organization. The Intel Select Solution[†] provides low latency and high throughput inference for deep learning networks. Further, these features, when integrated with the flexibility of the hardware of Intel, assist in preserving the investments. In the case of computer vision workloads, Intel has developed high-performance tools, open visual inference, and neural network optimization toolkits for deep learning deployments. The Intel Select solution applies

---

[†]Intel Select Solutions are pre-defined, workload, optimized solutions used to minimize the challenges related to deployment and evaluation of infrastructure.

the open visual inference and neural network toolkit for AI inferencing. The deep learning models trained from different networks are provided as input to the toolkit model, which helps to improve the flexibility of the processes. In order to train the deep learning models, high precision, 32-bit integers are transformed into 8-bit integers. Additionally, swapping the floating-point numbers helps to improve the accuracy through faster AI inferencing with the same accuracy, as the transformation of the integers. Moreover, the Intel Select solution also provides a solution for AI inferencing to multiple large models. The various Intel technology solutions are as follows:

(i.) Intel Distribution of OpenVINO Toolkit: The Intel Distribution toolkit is used for the analysis and adjustment of optimal inferencing deep learning models. These models are adjusted to provide smaller, faster, and 8-bit integer as an outcome with significant accuracy.

(ii.) Intel Math Kernel Library: To optimize the code for future generation Intel processors, the Intel math kernel library is used. This library is compatible with a wide range of compilers, languages, and operating systems.

(iii.) Intel Math Kernel Library for deep neural networks: This is an open-source library that improves the performance of deep learning networks associated with the Intel hardware.

(iv.) Intel AI frameworks for Python: There are different deep learning-based AI frameworks such as TensorFlow and ApacheMXNet. TensorFlow is a library that is easily and optimally applied to Intel Xeon processors. On the other hand, ApacheMXNet is an open-source deep learning framework for Intel MKL instructions.

(3) High-performance computing (HPC) clusters: Intel provides a solution for computationally intensive resources necessary to compute AI data on the existing HPC clusters. Intel Select solutions are the combined form of Intel Xeon Scalable processors, Intel high-performance computation architecture, and other technologies developed by Intel with batch schedulers. These solutions are applied to AI frameworks and are capable of performing computation on big data. The Intel solutions improve the performance and scalability of the framework. Additionally, the HPC clusters also satisfy industrial standards. Further, this architecture allows simulation and modeling, big data analytics, and AI on a common HPC platform. The customers are capable of executing AI workloads on the existing HPC clusters. The major challenges faced during the implementation of HPC clusters are the appropriate allocation of resources and determination of the pattern of computing systems based on GPUs.

Summary

1. The applications of AI and ML have resulted in the reformation of the various industrial processes.

2. To apply various ML algorithms, a significant amount of data is required. Based on the sample dataset, the training model is designed.

3. The different types of ML algorithms are – supervised, unsupervised, semi-supervised, and reinforcement learning. A labeled dataset is necessary for supervised learning. On the other hand, a limited set of the labeled and unlabeled dataset is used for semi-supervised and unsupervised learning.

4. In different industries, solutions/platforms, such as C3IoT, GE Predix, and Oracle, improve the efficiency, security, and upgrade the quality of the product through the application of ML.

5. To extract meaningful information from real-time data generated, the application of data science in industries helps to achieve proper business outcomes, increase system throughputs, improve product qualities, and identifies corrective measures to be taken.

6. Based on ANN, deep learning algorithms are developed. Typically, there are three layers in an ANN–input, output, and hidden. Synaptic weights interconnect these layers. Further, these weights are updated using the backpropagation method.

7. Feedforward neural network is the simplest form of neural network. A RNN is an ANN with the addition of a feedback loop to the feedforward neural network. In RNN, the weights are adjusted using backpropagation through time.

8. CNN is another form of a neural network, specifically applied for detecting hidden patterns from the analysis of data, especially images.

9. Deep learning is applied across various industrial sectors such as manufacturing, healthcare, business, and supply chains, to improve overall productivity, optimize prices, and enhance the flexibility of hardware.

Exercises

1. What are the various types of attributes of data? Give examples of each type in industry.

2. What is linear regression? Give a case study on application/s of linear regression in food industries.

3. For which type of applications is random forest used? Provide an e-commerce scenario to justify the application of a random forest.

4. How are different types of ML algorithms classified?

5. What are the applications of ML in C3IoT?

6. In which industries are the analytics platform developed by GE Predix is applied? Why?

7. What are the different advantages of data science in industries?

8. How are the weights updated in deep learning methods?

9. What are the common methods used for classification in deep learning?

10. How deep learning methods are applied in Toshiba? Discuss some of the AI analysis methods used by Toshiba.

11. Discuss the different features of H2O platform.

12. Describe the solutions developed by Intel for AI inferencing.

13. What are the major challenges faced in the implementation of high-performance clusters?

14. Write the differences between the CNN and RNN.

15. How do the analytics-based cloud services help the customers?

Applications and Case Studies

14

Healthcare Applications in Industries

Learning Outcomes

- This chapter covers the basics of IIoT-based healthcare applications.

- New readers will get an idea of the various IIoT-based healthcare systems, and smart devices used to monitor the health of a patient. Additionally, the advanced technologies used to improve the quality and minimize the costs of healthcare services are discussed.

- Seasoned learners will get an overview of the IIoT-based healthcare system and various smart devices used in healthcare.

- The future research directions in the domain of healthcare are also discussed in brief.

14.1 Introduction

In the last few decades, the integration of IoT has resulted in the all-round development of the healthcare sector. Previously, patients interacted with doctors through limited visits, tele- and text-based communications. With the advent of IoT-based sensors and advanced networking technologies, remote monitoring of the health of patients is now a very robust and economical solution. This has led to a reduction in the duration of patient's stay in the hospital. However, these technologies are not able to prevent the processes of aging and chronic diseases associated with humans. Affordable healthcare solutions are in high demand in the present day. With the application of IoT-based technologies in healthcare industries such as hospitals and research centers, the quality of services provided to the patients and the efficiency of the medical professionals have improved. One such technology, the Wireless

Body Area Networks (WBANs), comprises of set of sensor nodes, which sense the physiological parameters of the patient and transmits them to a remote server through a Local Processing Unit (LPU). In case of any criticality in the patient's health condition, an alert is transmitted to the remote monitoring unit. These physiological data of patients are prone to attack during transmission to clouds or their integration with new devices. The privacy and security threats associated with the physiological data of the patients is a major challenge to be addressed in healthcare systems.

14.1.1 Major challenges associated with healthcare

In the last few years, the healthcare industry has faced many challenges due to an increase in population, aging-related problems, an increase in various types of diseases, and a simultaneous increase in the demands of people/patients. Some of the biggest challenges in the healthcare industry are as follows:

- Avoidable medical errors: The commonly occurring medical errors are drug overdose, a wrong drug given to patients, the wrong combination of drugs, and adverse reactions to drugs. The appropriate medicine and proper dosage taken at the proper intervals, as suggested by the doctor, is important. The adverse reaction of patients to drugs may have minor or major after-effects on the patient. Further, drug overdose or an inappropriate combination may cause permanent damage to any organ or may even result in death. Similarly, as an example, the appropriate amount of oxygen supply to be provided to any newborn baby is to be maintained carefully by the medical experts. Overdose of oxygen supply may have toxic effects on the babies.

- Increasing age of population: The major cause of the increase in the aging population is the change in the causes leading to death, such as infections and chronic non-communicable diseases. The most frequently occurring age-related diseases are diabetes, arthritis, heart diseases, hypertension, and cancer. With the increase in the diverse form of diseases, certain challenges such as lack of availability of expert care-givers as per patient's disease, shortage of family care-givers, scarcity of healthcare professionals, and scarcity of resources are now crucial challenges in long-term healthcare.

- Costly hospitals and clinics: In the last few decades, there have been many new developments in the field of medical science. There are medicines and surgeries, which now promise a cure to complicated diseases, which was previously incurable. However, these medical solutions are quite expensive. The factors responsible for the rise in costs of drugs, pathological tests, and surgeries are the increase in population, rise in the aging population, type of care given to the patients, and increase in the chronic illnesses among various age groups.

- Steady rate of increase in research: The new findings in medical science take a long time for acceptance and to be available to human patients. Medical practitioners mostly suggest the medicines which are long-term in practice and given to patients over extended time-period. Therefore, the development of a new drug is risky, expensive, and a long-term process [193].

14.1.2 Coping with increase in diseases

With the rise in population, the type of chronic diseases affecting the patients has increased at an alarming rate. The medicines given to the patients are costly, and the consolidated treatment over extended periods is sometimes so expensive that only a select few can afford them. These medical expenses can be reduced through telecare applications, smart home, or telemedicine. The health condition of an elderly patient may fluctuate at any time instant. Therefore, continuous monitoring of these patients reduces the chances of admitting the patient to hospitals. WBANs comprise of multiple physiological sensor nodes which monitor the health conditions of the patient. These sensor nodes measure the physiological parameters of the patient, such as ECG, blood pressure, glucose levels, heart rate, and temperature. Further, the physiological data of the patients are remotely monitored by nurses/paramedics. In case of any abnormality in the patient's conditions, an alert is transmitted to the remote monitoring unit.

14.2 Applications of Healthcare in Industries

In the present era, smart and intelligent devices have been widely adopted in different healthcare applications. The increase in the aging population and the rate of increase in the diverse forms of chronic diseases cannot be reduced. However, the introduction of various wearable devices may minimize the cost and frequency of regular health check-ups. Additionally, the symptoms of chronic diseases may be detected at their onset, in the preliminary stages itself. Therefore, regular medical check-up of patients can be transformed from hospital-centric to home-based. The health conditions of patients can be remotely monitored through sensor nodes, which are either placed on or in the body of the patients. Healthcare professionals can access, review, monitor, and provide feedback to the patient based on the data transmitted by these sensor nodes. To ease the suffering of the patient, 'smart beds' have been introduced. A smart bed possesses different features, such as sleep tracking, temperature control, air chambers, app integration, and position control. Fig. 14.1 illustrates the schematic representation of the data flow in an IIoT-based healthcare system.

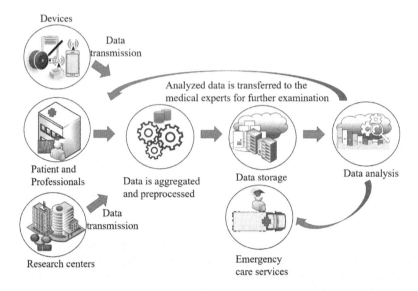

FIGURE 14.1: Schematic representation of data flow in an IIoT-based healthcare system [194].

Multiple smart and intelligent devices such as smartwatch and smart healthcare apps, sense and transmit data to servers through gateway nodes. The transmitted data is aggregated, processed, and stored at the cloud/server. Further, these processed data can be examined by medical experts and analyzed at research centers. The medical experts upload the examined reports through a Web portal. The prior intimation of the health conditions of any patient will help to initiate treatments and timely application of proper drugs. Prompt decisions can be taken by medical practitioners in real-time, to provide better treatment to their patients. In case of emergency services, a critical patient's data may be processed at the network edge to reduce latency* and reduce power consumption† [195]. In addition to this, the real-time alert, tracking, and monitoring of the patient's conditions help in improving the accuracy of the variation in physiological parameters of the patient, enables quick intervention by the doctors, and improves patient care [197].

---

*Based on the health criticality of the patient, the physiological data of the patient are processed at the network edge or cloud to avoid the delay incurred in transmission and processing of data.

†As the sensor nodes are energy-constrained, any unnecessary power consumed by the sensor nodes may be avoided. Moreover, doctors can remotely monitor critical patients and prescribe solutions. In the case of critical patients, the sensed data are processed at the network edge, which helps to reduce the power consumption of the sensor nodes.

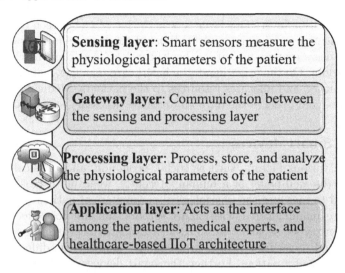

FIGURE 14.2: Block diagram of an IIoT-based healthcare system.

An IIoT-based healthcare system supports different technologies and is represented as a four-layered architecture – sensing, gateway, processing, and application layer. The functionalities of the four layers are as follows:

- Sensing layer: The sensing layer comprises of smart and intelligent devices, where the sensors are deployed or integrated into them. These sensors can measure various physiological parameters of the patient, such as ECG, body temperature, pulse rate, airflow, EMG, and oxygen saturation of the blood, as illustrated in Fig. 14.2. The sensor nodes are placed either on/into the body of the patients [196]. Moreover, these sensor nodes sense and transmit the data to the processing layer via the gateway layer.

- Gateway layer: This layer acts as the medium of communication between the sensing and processing layers. Additionally, this layer supports different protocols, which help in Machine-to-Machine (M2M) communication. These M2M networks may be public, private, or hybrid. Recently, several research works proposed that the physiological data of critical patients can be processed at the network edge to avoid further delays.

- Processing layer: This layer processes the various physiological data of the patients, stores, and analyzes them to extract meaningful information from them, as shown in Fig. 14.2. As a huge volume of data is transmitted by the sensors, automated analytics help in faster processing of the collected data. Moreover, the security of the data is maintained throughout the network. Typically, the cloud platforms store the massive amount of data generated from the sensing layer.

- Application layer: This layer acts as the interface between the end-users, medical practitioners, and IIoT-based healthcare infrastructure. The end-users request for various healthcare services from the cloud service provider. On the other hand, the healthcare manager manages the data, monitors, and allocates resources to the patients. Medical practitioners access and observe the data and provide necessary suggestions.

14.2.1 Smart devices

There are various factors, such as social, economic, and physical character-istics, for the well-being of a human. Recently, wearable fitness devices are being used to measure the physiological parameters of patients. These smart devices are manufactured with smart sensors and are cost-effective. There are different types of smart sensors, such as camera, proximity sensor, light detector, and fingerprint, which can remotely monitor the different physical characteristics of a patient. Further, the smart devices, as shown in Fig. 14.3, are used for personalized monitoring of health conditions such as diabetes, obesity, depression, and mental health. Some of the smart devices used for checking the health conditions of a patient are as follows:

- Smart Electrocardiogram (ECG) Monitor: The ECG signal of the pa-tient is first recorded through the connected devices and transmitted through the gateway. Thereafter, the recorded signal is reconstructed, enhanced, watermarked, and sent to the cloud. Fig. 14.4 shows the schematic representation of the IIoT-based ECG monitoring system. The cloud service provider collects these ECG signals, stores them, and extract important features. Further, the patient's ECG is transferred to the healthcare professionals for assessment. Finally, the reviewed ECG is uploaded to the IIoT-based ECG visualization and analytics platform. The results are then sent to the corresponding patients. Therefore, this system enables the doctors [200] to remotely monitors patients remotely. QardioCore is the world's first wireless ECG monitoring device available in the market, which tracks the health of the patient's heart and is easy to use. This device is useful for patients with a history of heart attacks or strokes, high blood pressure, high cholesterol, diabetes, and excess weight [198].

- Glucose Level Monitor: A glucose-level monitor helps to measure the glucose level of a diabetic patient. The sensor is usually placed under the skin, and the transmitter is attached above the sensor. As diabetes is a metabolic disease, therefore, monitoring the glucose level of a diabetic patient at regular intervals is essential. Based on the glucose level, the patients can plan their meals, physical activity, and medicines. Dexcom is a simple, user-friendly glucose-monitoring device usually utilized to monitor the glucose levels of patients. On the other hand, another form of wireless glucose level monitor, manufactured by i-Health, comprises

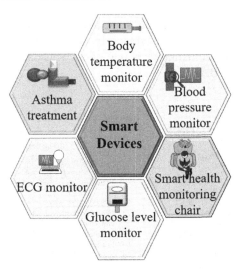

FIGURE 14.3: Smart devices for checking the health conditions of patients.

a detachable dongle. The blood sample of the patient is placed on the sample strip and slipped into the dongle. Thereafter, with the help of the registered app, the glucose level is determined.

- Blood Pressure Monitor: The blood pressure of the patients can be measured using the IIoT-based sphygmomanometers. Further, the blood pressure of the patients can be remotely monitored by the doctors in real-time. The Withings blood pressure monitoring device developed by i-Health monitors both the systolic and diastolic blood pressure of a patient. Further, the report is transmitted to the patient's mobile or a doctor through Bluetooth.

- Body Temperature Monitor: The body temperature of the patients can be continuously monitored using wearable sensors. Kinsa is one of the examples of the IoT-based body temperature monitoring device. This device takes into account the age, temperature, and symptoms of the patient to provide suggestions for the appropriate medicines or alert doctors. The suggestions are provided to the patient through smartphone-based applications.

- Oxygen Saturation Monitor: The term 'Oxygen Saturation' in blood indicates the ratio of the amount of oxygen-saturated hemoglobin to the total amount of hemoglobin[‡] in blood. The oxygen saturation level of a patient is monitored using the pulse oximeter. It continuously measures

---

[‡]Total hemoglobin = (saturated hemoglobin +unsaturated hemoglobin).

the blood oxygen saturation level non-invasively. Additionally, it is a low-cost device, which can remotely monitor the patient's health conditions. The pulse oximeter is one of the sensors integrated with other sensors in a WBAN.

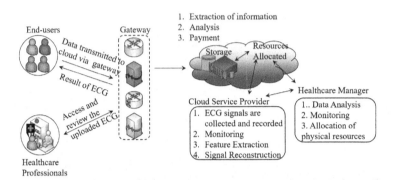

FIGURE 14.4: Schematic representation of ECG monitoring in IIoT scenario [194].

- Asthma Treatment: Asthma is a form of respiratory disease, which requires long-term treatment and cannot be cured completely. inhalers provide quick-relief and fast-acting medications for asthma. In continuation, smart inhalers can detect the timings when the inhaler is to be used and transmits data through Bluetooth devices paired with them. A smartphone can track and explain the data obtained from these inhalers.

- Smart Health Monitoring Chair: The health conditions of a patient can be continuously monitored through a smart chair [201]. With the help of non-intrusive methods, the electrocardiogram (ECG), photoplethysmogram (PPG),§ and ballistocardiogram (BCG)¶ measurements of any patient can be done simultaneously. Therefore, the heart rate and blood pressure of a person can be continuously monitored through this chair. ECG of the patient can be measured in the form of an electrical signal obtained through sensors attached to the user's clothes. This chair can be used to monitor the health conditions of aged persons and patients during regular activities.

---

§The instrument, which optically measures the changes in the blood volume of a person. Typically, a pulse oximeter may be used to measure the PPG of the patient.
¶This is a method of measuring the repetitive actions of the human body which emerge due to the unexpected ejection of blood into the vessels of the heart.

14.2.2   Advanced technologies used in healthcare

The integration of advanced technologies in the healthcare sector has resulted in improved patient monitoring, reduced costs, and minimized the waiting time for patients. The quality of healthcare services provided to the patients has become better, leading to saving millions of patients' lives [202]. Fig. 14.5 depicts the advanced technologies applied for healthcare applications, which are discussed as follows:

- Artificial Intelligence (AI): AI is widely applied across healthcare industries for scheduling appointment of doctors and suggesting hospitals, based on the severity of the health conditions of a patient. Additionally, depending upon the criticality of patients, they can be shifted from the Intensive Care Unit (ICU) to general wards, with minimum intervention of doctors. Further, considering their unique body chemistry, medicines can be recommended to the patients.

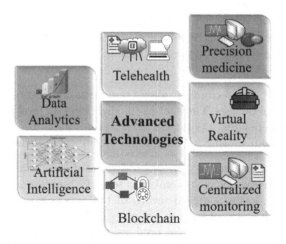

FIGURE 14.5: Advanced technologies used in IIoT healthcare scenarios.

- Data Analytics: The analysis of the collected data helps to provide meaningful insights and help medical experts/assistants in decision making. Through macro- and micro-level analysis of data, the overall costs associated with the medications of patients can be minimized. Further, the waiting time for the patients during their visit to the hospital or doctor's clinic can be reduced. Additionally, the quality of care given to improve patient satisfaction through the organization of the process of making appointments, insurance claims, and providing referrals are improved. Moreover, the health conditions of the patients can be predicted before the initiation of their treatment [199].

- Centralized monitoring of patients: Centralized monitoring facilities help technicians to remotely monitor the patients, which helps the medical experts to communicate with the caretakers directly. With the help of sensors, high-quality camera, and pulse oximeter, the conditions of heart, blood pressure, and respiratory rate of the patients can be monitored. In case of any emergencies, an alert is transmitted to caretakers as well as doctors at the remote end.

- Precision medicine: Precision medicine indicates the method of provisioning customized medicines, products, and medical practices, to individual patients. Further, based on the health conditions of the patient, the cost of medical decisions, practices, and products, are customized. For example, in the case of cancer patients, based on the genes of the patients, targeted treatment is done to induce gene mutations.

- Virtual reality (VR): Virtual reality-based technologies help in the diagnosis of diseases, treatment of patients, and their recovery. Medical students get real-life training experiences such as visual understanding of human anatomy, and immersive experience of the different surgical procedures, through virtual reality devices. In addition to this, VR technologies help in the treatment of phobia, robotic surgery, depression, and blood loss of patients.

- Telehealth: This technology allows remote monitoring of patients through their digital devices. The doctors/medical experts provide suggestions to patients depending upon their health conditions. The patients are also relieved from waiting for long hours during their visits to the doctor.

- Blockchain: When the physiological data of the patients are transmitted to the server, it is vulnerable to security breaches and manipulation. Blockchain keeps a digital record of the sensitive medical data of patients in the form of ledgers, which is difficult to tamper. Additionally, these blocks of data are secured using cryptographic principles. This blockchain-enabled healthcare system maintains a distributed record of the medical data, which helps the doctors to access these data, identify the disease, and provide a proper diagnosis.

14.2.3   Open research issues to be addressed

We discussed an IIoT-based healthcare system, various smart devices used, and advanced technologies applied to provide improved healthcare facilities to the patients. However, there are various research challenges to be addressed in the future, which are as follows:

- The improvement in the scalability of healthcare systems is required without compromising the security of the system. The medical data of the patients are prone to attacks; therefore, these data should not be accessible to unauthorized users, and any form of modification of the data is undesirable.

- As the patients may be mobile, the WBAN system may be upgraded and extended for mobile devices and environments.

- As the sensor nodes are energy-constrained, IIoT-based healthcare technologies, methodologies, and protocols may be designed in such a way that energy consumption is reduced.

- The computational capability of healthcare devices requires improvement to enable complex operations.

- Reducing the amount of memory utilized to process and analyze the sensed physiological parameters of the patient.

Summary

1. The integration of advanced technologies with the traditional healthcare system has resulted in minimizing the cost of healthcare facilities, remote monitoring, and improved the security and privacy of the physiological data of patients.

2. The major challenges faced by the healthcare sector are – commonly occurring medical errors such as drug overdose and wrong combination of drugs, increase in the aging population, hike in expenses of drugs, pathological tests, and various costs in hospitals, and steady rate of research in medical science.

3. Smart and intelligent devices are widely used in different healthcare applications such that patients can be remotely monitored, reviewed, accessed, and healthcare experts provide suggestions.

4. IIoT-based healthcare system comprises of four layers – sensing, gateway, processing, and application.

5. Various smart devices are used for checking the health conditions of a patient, such as smart electrocardiogram monitor, smart glucose level monitor, blood pressure monitor, body temperature monitor, oxygen saturation monitor, asthma treatment, and smart health monitoring chair.

6. In order to improve the healthcare facilities, advanced technologies such as artificial intelligence (AI), data analytics, centralized monitoring of patients, precision medicine, telehealth, blockchain,

> and virtual reality (VR)-based technologies, are integrated with the existing healthcare system to upgrade the scope and quality of services.
>
> 7. Various research challenges to be addressed in this domain are scalability, energy-constrained nature, computation capability of devices, and amount of memory utilized.

Exercises

1. What are the commonly occurred medical errors?

2. Discuss the architecture of the IIoT-based healthcare system.

3. What are 'smart beds'? What are the features of a smart bed?

4. Describe the working of a smart ECG monitoring system in the IIoT scenario.

5. What are the features of a smart health monitoring chair?

6. Discuss the importance of blockchain in healthcare industries.

7. What are the applications of precision medicine in healthcare?

8. How are patients monitored centrally in a hospital scenario?

9. Write some of the research challenges to be addressed in the future to improve healthcare facilities.

10. How does a glucose level monitor works?

15

Inventory Management and Quality Control

Learning Outcomes

- This chapter outlines the different applications of IIoT in inventory management and quality control.

- New readers will be able to visualize the various types of inventory, types of inventory management methods, and applications of the IIoT-based technologies for inventory management.

- For seasoned learners, this chapter will help to brush-up their concepts on inventory management using IIoT, its benefits, and provide insights into quality control.

15.1 Introduction

The raw materials, finished goods, and assets form the inventory of an organization. Inventory management is the process of management, storing, ordering, warehousing, and processing of inventory. To avoid shortages and fulfill the demand of customers, proper management of inventory is essential. In the case of companies with complex supply chains and manufacturing processes, appropriate inventory management is vital to keep track of the finished goods, amount of remaining raw materials, and finding the probable time for re-ordering inventory. The integration of advanced technologies with traditional inventory management processes may minimize the total cost of inventory, errors involved with the stock review process, and fulfill customers' demands within the stipulated time. Smart sensors and Radio Frequency Identification (RFID) tags attached to machines and products help to track the items in real-time. The data sensed by these deployed sensor nodes are transmitted to the cloud for storage, processing, and analysis. Further, the real-time update of the inventory's location helps to monitor inventory. The amount of inventory required in the forthcoming industrial processes can be forecasted, and meaningful insights drawn from them from the analysis of collected data. Therefore, the manufacturers can jointly optimize the amount of on-hand inventory to be maintained and new inventory to be added to fulfill the customers' requirements in the least amount of time. Moreover, the obstruction caused in the manufacturing processes due to improper utilization of resources and the unavailability of raw materials can be eliminated to a certain extent through inventory management and control. Additionally, the lead time* required for inventory management is reduced with the help of IIoT-based technologies.

15.2   Inventory Management

The end-to-end process of receiving raw materials, manufacturing goods, transportation of finished goods, and warehousing is termed as inventory management. Fig. 15.1 shows the schematic view of the inventory management process in a plant. The mapping from man-to-goods to goods-to-man inside a warehouse is the transformation of a traditional process to IIoT-based inventory management. This metamorphosis in the inventory management system helps to improve efficiency, minimize costs, increase overall productivity, and conserve energy. Moreover, appropriate inventory management maximizes the available space in the warehouses, optimizes the workforce required, and increases customers' satisfaction.

In the Amazon Warehouse, about 1 million inventory of different types is autonomously maintained with minimum human involvement, while improving and maintaining quality-based attributes. By the second quarter of 2019, Amazon fulfilled customers' demands and attained a profit of USD 800 million. However, as the time for delivery of goods to the customers is reducing, the transportation costs keep increasing. In order to reduce the transportation cost, Amazon bought around 50 airplanes, 300 semi-trucks, and 20,000 delivery vans. They employed supersonic conveyors to meet the expectations of customers. The speed of these conveyors varies with the flow order levels. Further, various types of robots, such as driving robots to move inventory, robotic arms for lifting boxes, and robots for customized packaging, are used in the warehouse. Therefore, with the integration of IIoT-based technologies, Amazon has become one of the biggest global retailers in the last 25 years [204].

---

*Time taken from initiation to delivery of the product is known as lead time. Based on the re-ordering point, the lead time may vary.

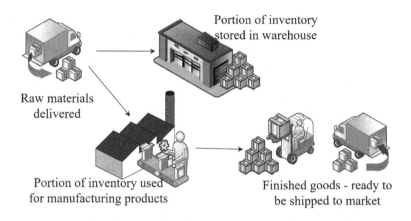

Portion of inventory stored in warehouse

Raw materials delivered

Portion of inventory used for manufacturing products

Finished goods - ready to be shipped to market

FIGURE 15.1: Schematic view of inventory management [203].

15.2.1 Inventory

Inventory is the stock of goods or products available for sale and raw materials utilized for production. It is also classified as the asset of an organization. The three basic metrics of inventory are *level, turnover ratio*, and *cycle time*. Practically, there are different types of products/items/goods stocked in the store. The levels of these inventories are checked either quarterly/monthly to avoid stock-out. These inventories are converted to usable form or sold in the same form as they arrived. Therefore, the turnover ratio of inventory may vary from weeks to months. Further, the reorder point needs to be computed to ensure the smooth running of processes. The time required to complete the process, referred to as its cycle time, is vital. Based on the cycle time, new orders managed, and old orders fulfilled. In an automated process, tracking the levels, ratios, and cycle time can be updated dynamically (Fig. 15.2). The various types of inventory are as follows:

- Finished goods: The finished goods are referred to as the inventory of completed products. The final inspections regarding the quality of the product are done before transferring the product from the work-in-progress section to the completed goods section. The finished goods inventory can be classified as:

  1. Transit inventory: The finished products which need to be transferred to a location from its present location are known as transit inventory. It is also known as pipeline inventory. The transport of finished products from regional warehouses to state warehouse or warehouse to retailers may take days or even weeks to reach the destination through freights. However, the combination of the inventory and shifting it may minimize the time taken to transport these inventories.

2. Buffer inventory: As the name suggests, the inventory is used to insulate a supplier from the occurrence of unusual events such as poor quality/damaged products delivered, uncertainty of supply and demand, and stock-outs, within the industry. The buffer inventory is also known as safety stock. With the increase in buffer inventory, the possibilities of stock-outs reduce. As a result of the reduction in stock-outs, orders are fulfilled within the stipulated time, and the likelihood of customers' satisfaction increases.

3. Anticipation inventory: Based on their prediction regarding a future event such as Christmas or regional festivals, the appropriate surplus amount of inventory is held by the manufacturers, which is referred to as anticipation inventory. As the seasonal demand for certain products rises, the price of these products increases. The possible reason for the hike in prices may be the unavailability of skilled laborers and excess overtime costs demanded by them. However, when demand is low, workers can be kept busy by increasing the production time. This helps the manufacturers to maintain a constant level of output and a stable workforce. Therefore, the process, as mentioned earlier, helps to reduce spikes in the demand curve and is known as smoothing.

4. Decoupling inventory: In any industrial process, most of the associated machines may possess matched production rates. However, in case multiple machines are under repair or undergoing preventive maintenance, the workflow should not be interrupted. This form of inventory is termed as a shock absorber, which protects the system against irregularities in production. Therefore, the dependency upon the factory's unit reduces to a certain extent. An optimum balance can be maintained between the inventory levels through better coordination among the systems. Higher the amount of decoupling inventory in any plant, lower are the dependencies for the successful operation of the plant.

5. Cycle inventory: The inventory holding costs increases when large quantities are ordered/manufactured despite the reduction in ordering/setup costs. When the lot sizes ordered are reduced, the inventory holding costs also reduce. As a result, the ordering/setup costs increases to meet the demand. The total cost can be minimized when the ordering/set up costs are equal to the inventory holding costs. Based on the daily demand and lead time, the manufacturer's reorder point and the amount of safety stock to be maintained are computed. The solution to this problem is fixed quantity, known as Economic Order Quantity (EOQ), to be ordered to minimize the annual cost for retaining inventory. EOQ retains a balance between the inventory holding costs and ordering/setup costs. Cycle inventory results from ordering the items in batches, or lot sizes.

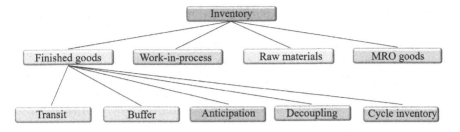

FIGURE 15.2: Types of inventory.

- Work-in-process: Work-in-process inventory comprises the materials, processes, inventories required to manufacture the assemblies and sub- assemblies, which are processed or waiting to be processed in the system. The materials included in work-in-process inventory comprise of raw materials provided for the initial set up of processes and materials, which are completely processed and waiting to be inspected and accepted as finished goods.

- Raw materials: The inventory items used by the manufacturer to produce assemblies, sub-assemblies, and products, are known as raw materials. These inventory items may include extracted materials to be produced, and objects or elements that have been externally purchased by the organization. Further, these inventories also comprise of partially assembled goods or ones considered as finished by the manufacturer. For example, grains, minerals, chemicals, wood, paint, and steel are the raw materials of an organization. Additionally, nuts/bolts, ball bearings, seats, and wheels, which are purchased from outside firms, are also raw materials. Typically, raw materials are utilized to manufacture a component. These manufactured components are placed into sub-assemblies, which are fit together to produce the final product.

- MRO goods: The maintenance, repair, and operating supplies, or MRO goods form the basic support of industries and maintain the industrial processes. The inventory which is involved in the manufacturing process and is indirectly related to the finished goods forms the MRO goods. For example, nuts, bolts, lubricating oils, screws, gloves, uniforms, and coolants are some of the MRO goods.

15.2.2   Types of inventory management

Based on the type of business or products analyzed, inventory management methods are categorized as Just-In-Time (JIT) management, Materials Requirement Planning (MRP), EOQ, and Days Sales Inventory (DSI).

- JIT Management: This concept was introduced in Japan during the period between the 1960s and 1970s. The primary aim of JIT was to improve overall efficiencies and reduce the number of wastes and costs associated with the manufacturing process. Additionally, the demand could be forecasted with higher accuracy to provide insights into the customers' behavior. However, in case a specific item goes out-of-stock, there may be risks of not being able to fulfill the orders on time.

- MRP: MRP is one of the most widely used systems applied to compute the number of materials and components required to manufacture a product. Therefore, the forecast of the amount of sales should be done by the manufacturers within the allowable time limits such that the required inventory is planned and timely ordered to suppliers.

- EOQ: This method of inventory management is applied to compute the optimal number of items ordered with each batch of order such that the total cost of inventory is minimized. The total inventory costs decrease when both the setup costs and inventory holding costs are reduced. Further, demand is assumed to be constant and known. Therefore, EOQ is independent of demand.

- DSI: The number of days or time required to sell the inventory, including the in-progress products, are known as DSI. It also represents the number of days after which the inventory requires reordering.

15.3 Inventory Management and IIoT

The integration of smart sensors and advanced technologies results in interconnected devices/machines and humans, allows for improved tracking of products, and fulfills customers' requirements in real-time. Further, real-time analysis of the huge volume of generated data helps to forecast the demand well in advance. Typically, in the IIoT scenario, the products, raw materials, or goods possess RFID tags.[†] Each of these RFID tags has a unique ID. As the RFID tags attached to the product are scanned, the IDs are extracted, and the collected data are transmitted to the cloud/server for further processing. This allows for the details of inventory items to be updated in real-time. The users are capable of monitoring these inventory in real-time, remotely. With the help of RFID tags, multiple items can be scanned simultaneously. The inventory overstocking or stock-out problems can be subsequently avoided.

[†]A RFID tag comprises a chip, an antenna, and readers.

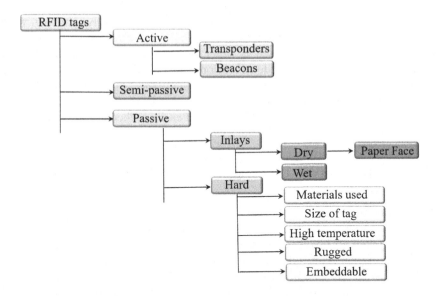

FIGURE 15.3: Types of RFID tags [205].

Additionally, the quality of the stored product and product security can be improved. Fig. 15.3 illustrates the various types of RFID tags.

An RFID tag conveys information, and the antenna provides the necessary power to transmit the information. Further, the readers present in the RFID tag are either fixed or handheld, which utilizes radio waves to write and read from tags on items. Usually, RFID tags are of three types – active, semi-passive, and passive. Active tag possesses the power source required to transmit data and is used for real-time tracking of assets. Moreover, active tags possess longer range and are expensive compared to passive tags. Active RFID tags are classified as transponders and beacons. Transponder tags first send a signal to the reader. The reader responds to the signal providing related information. On the other hand, the beacon tags do not wait for the response from the readers. Active beacon tags are specifically used in oil and gas industries for mining and cargo tracking. Semi-passive RFID tags possess an on-board power circuit. However, there is no active transmitter present in these tags. In addition to this, these tags do not depend upon the signals from the reader. Further, Passive RFID tags require external power to transmit information, which is generally provided by the tag readers themselves. Passive tags find common use in access control, file tracking, and supply chain management. Further, passive tags are small in size, cheaper, flexible, and persist for a long duration without batteries. Although various types of passive tags are present in the market, broadly passive tags are classified as – inlays and hard tags. RFID inlays possess antennae made of aluminum or copper and an integrated circuit (IC). Based on the substrate, the RFID inlays are categorized as wet inlays or dry inlays. Wet inlays possess an adhesive backing, while dry inlays appear laminated and possess no adhesive. Further, dry inlays with white paper or poly face, known as paper face tags, are mostly applied to read printed numbers or logos for identification. Hard RFID tags are manufactured from ceramic, rubber, plastic, or metal. Further, based on the temperature resistance of materials, rug-based material, size of the tag, and materials used in the tag, hard RFID tags are classified and easily embedded into the devices [205].

15.3.1   Benefits of IIoT applications in inventory management

The various IIoT-based applications in the inventory management system have resulted in the digitalization of manufacturing as well as business systems. For example, when an inventory moves from one unit to another, the RFID tags locate the items, scan, and transmit the detailed information regarding the item to the server. The inventory manager [206] can remotely view this real-time information about the inventory. The benefits of the integration of IIoT in inventory management are as follows:

- Automation in the process of inventory tracking and packing: With the integration of RFIDs and other IIoT-based smart sensors, the inventory can be tracked in real-time, as shown in Fig. 15.4. Further, these data are stored in databases at remote servers. Therefore, human efforts are minimized, the probability of human errors reduce, and real-time information of the current inventory stock is maintained.

- Real-time update of the quantity of inventory items, their location, and tracking: The improved inventory tracking system provides real-time information on the finished inventory, raw materials left in stock, work-in-progress inventory, to the inventory manager. Additionally, the inventory manager is also updated with the status, location, and movement of inventory inside a warehouse.

- Optimization of the amount of inventory: Real-time updates of the amount of inventory in stock help the inventory manager to regulate the inventory items in the right location, right time, and right quantities, as illustrated in Fig. 15.4.

- Recognize the reasons behind bottlenecks in operation: As the inventory manager get the real-time information of the inventory location and status, the machines with reduced operational efficiency, and amount of buffer and anticipatory inventory to be maintained, is identified by the manufacturer.

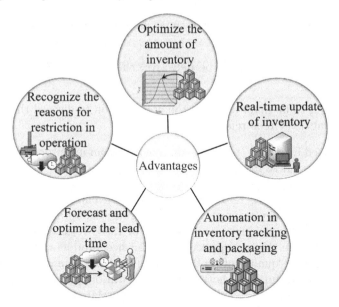

FIGURE 15.4: Benefits of IIoT-based inventory management.

- Forecast and optimize the lead time: The inventory manager possesses the data regarding the amount of inventory available and is able to forecast the buffer inventory in the warehouse.

15.4 Quality Control

Quality control (QC) is one of the vital aspects to be maintained across all industries. The step-by-step execution of processes and procedures to measure the quality of products (hardware and software components) through the evaluation of results is termed as QC. The quality of products is measured by preserving both the level and quality of output. In the case of inventory management, the quality of products and stored inventory are maintained and managed by the inventory manager. The inventory controlled software assists in controlling the quality of products during each of the stages of their development. The quality of finished products is measured through several stages of product testing. In case a product does not pass through the final stages, it is rejected. In addition to this, stock inventories also pass through several test stages to ensure the quality of the inventory. The rejected product/inventory which does not pass the QC are:

- The item under this category may be reworked or repaired through a fresh service order.

- If the item is found to be defective, the inventory is returned to the supplier.

- A list is to be maintained by the organization to discard these products and label them as rejected.

These rejected inventory/products are removed from inventory/finished goods count. Further, these items are kept aside in the warehouse as rejected stock. On the other hand, in the case of untested finished products, the product is labeled as 'on-hold'. Therefore, the IIoT-based inventory management methods help the inventory manager to track the rejected, approved items, reasons for rejecting these products, and location of these rejected products in the warehouse. Before the final packaging of finished products, it is very important to ensure the quality of products, such that the demand of customers is maintained in the future. Another different form of QC test complies with a specific environment, standard, and certificates related to any specific industry. However, each of the industries possesses certain internal QC metrics. The maintenance of QC metrics helps an organization to retain their reputation.

In order to maintain inventory quality, a new inventory management system may be developed to improve quality levels, minimize stock-outs, and reduce the amount of wastes produced. Advanced analytics may be applied to compute the re-order point, increase turn-over ratio, and upgrade data accuracy. Additionally, the probability of equipment or machine failure should be taken into account. The employees should be given specialized training to repair or replace equipment parts and handle situations emerging on the plant floor. Proper inventory management and QC help an organization to improve its efficiency, increase its profits in business, and provides a quality product to the customers. Therefore, satisfied customers form the repeat customers of the organization and provide another added advantage of positive feedback regarding the product.

Summary

1. The process of receiving raw materials, manufacturing goods, transportation of finished goods, and warehousing those products is termed as inventory management.

2. Depending upon the type of business or products analyzed, inventory management procedures are as follows – JIT management, MRP, EOQ, and DSI.

3. Inventory is the stock of goods or products available for sale and raw materials utilized for production.

4. There are different types of inventory – finished goods, work-in-progress, raw materials, and MRO goods. The finished goods can be further categorized as transit, buffer, anticipation, decoupling, and cycle inventory.

5. RFID tags are attached to the products, raw materials, or goods, to monitor, track remotely, and update the inventory in real-time.

6. RFID tags are classified as active, semi-passive, and passive tags.

7. The application of RFID tags help to automate the process of inventory tracking and packaging, real-time update of the inventory, optimization of the amount of inventory and forecast of the lead time.

8. QC of products is one of the necessary elements to be maintained across different industries. The rejected products which do not pass through QC are reworked or repaired, returned back to the supplier, and labeled as rejected.

Exercises

1. What is inventory management? What is the difference between a buffer and an anticipatory inventory?

2. Which inventory are referred to as work-in-progress?

3. How are IIoT applications for the management of inventory?

4. Write the differences between active and passive RFID tags.

5. Why are advanced technologies integrated into the system for inventory management in the field of IIoT?

6. What are the measures are undertaken, if the product does not pass through QC?

16

Plant Safety and Security

Learning Outcomes

■ In this chapter, we outline the applications of IIoT for plant safety and security.

■ New readers will be able to visualize the necessity of safety of workers at different locations in the workplace, and various IIoT applications applied to adopt safety measures.

■ Seasoned learners can brush up their concepts on software, network, and mobile device security.

16.1 Introduction

The safety of working professionals and machines at factory sites is an ever-present aspect of concern across different industrial sectors. Both health and safety are two key factors to be considered in the workplace to ensure the physical and mental wellness of workers and employees. Occupational and Health Safety (OSH) standards lay down certain rules and regulations to maintain health and safety standards, minimize injuries and fatalities of workers, improve the workplace environment, upgrade overall efficiency, and thereby improve product quality. The incorporation of various advanced technologies such as big data, advanced analytics, augmented reality, virtual reality, and 3D printing has resulted in the improvement of plant safety and security through the continuous observance of the different KPIs.* These technologies help to provide real-time information and understanding of the safety standards in

---

*Key Performance Indicators (KPIs) represent the performance measure to be undertaken by an organization to estimate their success. The number of near-misses, fatalities, ratio of inventory to sales, and the number of repeat customers.

factories. In addition to this, the health conditions of the workers, especially those working in hazardous zones, can be constantly monitored using wearable sensor nodes. Therefore, the various health metrics of workers, such as fatigue, stress, heart rate, and movement, are available in real-time. Based on their health conditions, these workers can be remotely monitored, and accidents can be avoided in the workplace. For example, in the case of an emergency, any worker engaged in risky and hazardous underground mines can be easily located and evacuated.

Workplace security is another challenge in industries. As multiple sensors are deployed at various locations and into the machines on the shop floor, an enormous amount of data is generated at every time instant. These data are prone to different forms of malicious attacks, which may cause a threat to the security of the organization itself. With the help of definitely described security policies, security can be improved in the workplace. Further, closed-loop security control leads to the minimization of malware threats. The entire security system should be updated and upgraded at regular intervals.

16.2 Plant Safety

Turbines, generators, pumps, and other types of equipment are considered as the assets of industries. Proper safety measures should be taken to protect these assets and workers engaged in the workplace at specific locations. Typically, various hazards are associated with installing, erecting, commissioning, decommissioning, and dismantling of machines or entire units in a plant. Plant safety systems are designed to detect any form of abnormality in the system, provide information to the workers, and stop operations to prevent accidents. To maximize workplace safety, certain safety measures are to be taken, which are as follows:

- Personal safety equipment is given to the workers and replaced after certain intervals.

- Operation manuals for different machines are created and distributed among the concerned workers.

- Initiation and application of strict administrative controls are required to handle hazardous materials in plants.

- Organizing safety training for individuals at regular intervals is crucial.

- Design policies to eradicate risks associated with processes or machines should be followed.

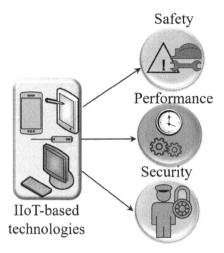

Safety

Performance

Security

IIoT-based
technologies

FIGURE 16.1: Safety aspects of an organization.

- Alcohol and drug testing programs can be arranged in each unit of a plant.

As an example, consider sensors are put on hard hats, gloves, uniforms, and biometric machines are installed at different locations inside the factory. These systems can help to identify a worker falling from a height. On the other hand, if any worker moves close to a machine, it is automatically shut down for a temporary period to avoid accidents. Moreover, based on the analysis of the data collected from sensors, accident-prone zones can be marked, and a safety alert can be provided to the workers. Therefore, the collected data helps to improve safety, enhance security, minimize downtime of machines, and assess the safety of the entire system. The integration of IIoT-based technologies helps in the improvement of safety, system performance, and security, as illustrated in Fig. 16.1.

Safety is an essential driver in improving the performance, efficiency, and operations of a plant. In many organizations, the Environment, Health, and Safety (EHS) professionals work together to maintain safety and eliminate the risks involved with the operations associated with these organizations. Operational excellence in an organization assists in forming close relations among energy consumption, EHS, quality, and application performance management (APM). These operational excellence parameters, as shown in Fig. 16.2, form the pillars of safety in any organization. The abnormal functioning of one of these pillars may lead to instability; abnormalities in the functioning of two or more pillars makes the entire system unstable. However, a closed-loop process of risk identification, assessment, control, monitoring, and response can lead to the minimization of risks, and thereby, improve safety and security in the industrial processes.

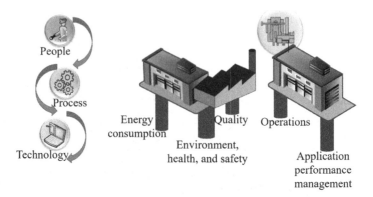

FIGURE 16.2: Pillars of safety in an organization.

16.2.1 IIoT applications for undertaking safety measures in plant

Smart sensors, AR- and VR-based applications, different protective types of equipment, and robots are applied across different industries to improve the safety of industrial assets and the human workforce engaged with them [207]. Fig. 16.1 illustrates the various aspects of an organization in terms of safety, security, and system performance. Some of the specific applications are as follows:

- IoT-based sensors and computer vision applications are integrated with traditional systems to ensure the safety of workers, extract information from the real-time collected data, discover defective products, locate the abnormalities in the system, and take necessary actions.

- The protective types of equipment used by workers such as safety goggles and high-visibility vests help to ensure their safety at the workplace. These equipment improve their visibility, monitor their postures, track noise levels on the factory floor, and provide real-time feedback to the workers.

- Integrated safety systems are designed to identify hazardous zones and install proper power-factor controllers in the industries.

- The primary focus of IIoT-based industrial systems is on security and malware threats, which may affect the connected equipment and compromise the safety of the entire system.

- Robots are introduced to perform complex operations. As humans and robots work together, safety improves and efficiency increases. A safe distance should be maintained between humans and machines to avoid undue incidents.

- Location-based analytics and real-time information collected from sensors improve the safety of assets and individuals. Additionally, warnings are given to workers in case they enter hazardous zones without appropriate safety measures.

16.3   Plant Security

IIoT-based technologies help plant operators to remotely access pump stations, turbines, or integrate any equipment present on the factory floor with other types of equipment. This results in the improvement of productivity, product quality, and efficiency. However, this also poses security threats, increases chances of damage to equipment, or raises safety issues for the workers. Security threats also arise in industries where men and machines work collaboratively. For example, in the case of chemical industries, the health risks are quite high for workers because the employees are exposed to volatile chemical materials. Moreover, a compromised system may pose a security threat to the customers as well. For example, any form of alteration or contamination in the products of a food or pharmaceutical industry can have toxic effects and prove to be life-threatening for humans. Further, if a lot of finished drugs are detected with abnormal behavioral trends, the delivery of the new lot of those life-saving drugs to the market will be delayed. However, to protect machines, workers, consumers, and intellectual property, the aspects of safety and security of the organization can be merged. As illustrated in Fig. 16.3(a), the three main pillars of an organization are culture, compliance, and capital, which are discussed as follows:

- Culture: The attitude, values, ethics, and priorities of the working personnel determine the safety and security policies of the organization. The cooperation among workers, IT professionals, and executives, working together towards protecting the assets and intellectual property of an organization is essential. Further, when each of the employees works together, the critical safety-related data and any form of abnormality in the processes or machines can be easily and timely detected.

- Compliance: The organization needs to abide by certain policies and procedures to maintain safety standards. Security efforts should follow the standard protocols and manage the security risks associated with individual organizations.

- Capital: An organization possesses certain in-built security and safety features to protect the assets and workers. In order to incorporate new safety and security technologies, companies need to optimize productivity and safety within their establishments. Further, the security measures

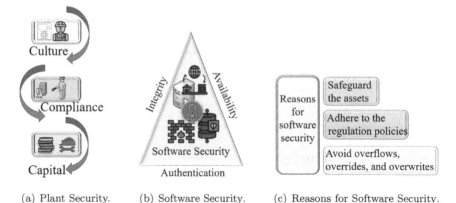

(a) Plant Security.   (b) Software Security.   (c) Reasons for Software Security.

FIGURE 16.3: Security.

used to protect the system against malware attacks and breaches should also facilitate and provide a speedy recovery of the infected system.

16.3.1   Software security

Software security is one of the important aspects, which protects software against malicious attacks in the industries. To skim information, attackers may monitor data streams and introduce malware into the systems. These may result in system crashes. In many instances, buffer and stack overflow attacks may lead to overwriting of contents. In case the computers are not kept on the same network, the attacks can be avoided; however, this solution seems to be quite impractical in a majority of industrial settings. Some of the prime reasons to provide software security are illustrated in Fig. 16.3(c) and are as follows:

- Safeguard the assets of the organization from malicious attacks: the essential information/data present in the networked computers form the assets of an organization. Any form of a malicious attack on these data may result in a violation of various security aspects of the industries. Typically, network and computer security involve protection, integrity, and availability of information.

- Adhere to regulatory policies and maintain the trust of the organization: certain rules may be maintained to avoid unrestricted access to the data. There is a high chance of shareholders investing in organizations, which secure and maintain the trustworthiness of data, such that the shareholders' investments are protected and secured. Therefore, the overall profit of the organization increases.

- Avoid overflows, overrides, and overwrites of the information/data of the organization: these forms of attacks are quite common and may result in the system to crash, besides manipulation of crucial data. This measure also protects against accidental manipulation or corruption of data.

Therefore, in order to minimize unauthorized attacks on networked computers, certain preventive measures should be undertaken. Proper policies and procedures such as backups, media control, disaster management, and contingency planning help to protect information/data in industries. However, software security is not absolute. The security requirements of the organization should be upgraded with time. Fig. 16.3(b) shows the three essential components of software security, which are–integrity, availability, and authentication.

- Integrity: In order to avoid unauthorized access, data consistency, accuracy, and trustworthiness should be maintained during the entire life cycle. Moreover, data should not be altered illegally during transmission. To ensure that data is not modified, certain files and user access permission controls are designed by the organization. In addition to this, some backup files must be available to re-establish the affected data to its original state.

- Availability: Typically, the availability of a system is verified in terms of hardware maintenance and performing repairs, such that the system is operational with respect to total time. The scope of availability in terms of software is affected when the failure of the component/s coincide. We can observe that a strong relationship exists between software availability and reliability.[†] Further, to restrain data loss due to any unpredictable events such as natural calamities and fires, backups may be maintained by the organization.

- Authentication: Authentication denotes the identification of the user through online portals. These portals act as the interface between the system and the user. A user can access the system only after verification of their credentials. Therefore, the resources and assets of an organization are protected based on these access control methods.

In order to obtain software security, proper programming techniques should be used, updated software should be deployed for protection, any form of intrusion into the system should be identified, and preventive measures should be applied.

[†]Software reliability is defined in terms of the probability that the software operates properly in a specified environment within a stipulated time duration.

16.3.2 Network security

The event of sharing information and services among the interconnected devices to secure them is termed as network security. The primary objective of network security is to detect the various types of threats and prevent them from entering the concerned network. Multiple layers of defense mechanisms combine at the edge and in the network to secure the devices connected to it, where the rules and regulations for each layer [209] are implemented. Fig. 16.4 shows the various types of network security, which are discussed as follows:

- Access control: To avoid unauthorized access, a restricted group of users and devices are selected, which also need prior identification. Non- compliant users are blocked or listed under restricted access.

- Antivirus and anti-malware software: Some of the malicious software such as viruses, worms, Trojans, and spyware, may infect and remain inert for an extended time duration. Continuous tracking of files eliminate the detected malware from the system and affix the damage caused to the system.

- Application security: To preserve applications in a system from malicious attacks, holistic protection of the entire system is necessary. Practically, regular applications may comprises of security loopholes through which attackers may intrude into the network. Application security includes the security of hardware, software, and processes to avoid these holes.

- Behavioral analytics: Any form of abnormality in the network behavior can be detected by applying advanced analytics tools. Behavioral analytical tools can easily detect this form of malfunctioning of the network and restore the regular operation of the system.

- Data loss prevention (DLP): The working personnel in an organization may transmit sensitive information outside the network. To avoid loss of crucial information, DLP technologies may be used by an organization. These technologies restrict people from uploading, forwarding, or printing critical information from the organization's systems.

- Email security: One of the greatest security threats are email gateways. Attackers use personal information and social engineering strategies to redirect email recipients to sites laden with malware. Email security applications block incoming attacks and restrict outbound messages to prevent loss of sensitive information.

- Firewalls: Firewalls act as a fence between the trusted internal network and unauthorized external network. A set of rules are set by firewalls to allow or block traffic. Additionally, firewalls can be in the form of hardware, software, or both.

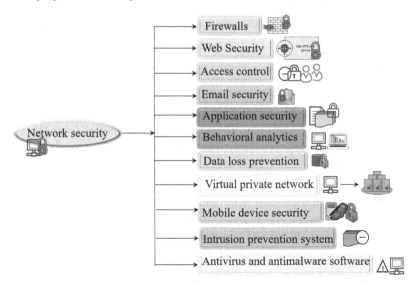

FIGURE 16.4: Types of Network Security.

- Intrusion prevention systems (IPS): The IPS examine the traffic and block possible active attacks. For example, CISCO next Generation IPS appliances block intruder attacks and track the infected files and the presence of malware in the network.

- Mobile device security: With the increased corporate applications on mobile phones, cyber-attacks on personal mobile devices have also increased. To avoid unauthorized access, devices that access the organization's network should be controlled.

- Virtual Private Network (VPN): VPN is a programming environment that produces a safe and encrypted connection from endpoints to the network.

- Web Security: Web Security solution controls web-based threats, access of users, and disallow the access to malicious threats. These solutions secure the web gateway either on-site or in the cloud. Further, web security denotes the stages to secure the entire system.

16.3.3 Mobile device security

In the past few years, most of the mobile devices such as smartphones, tablets, and laptops are being popularly utilized for various business applications. With the increase in the number of mobile devices and corporate applications executed through them, these devices have become more prone to malicious attacks. However, these mobile devices can be secured using a multilayered security approach [210]. The various components to secure mobile devices are as follows:

- Endpoint security: To ensure endpoint security, organizations should be capable of examining the files and processes on each of the mobile devices. These flexible and mobile workforces can be remotely accessed. Malicious behavior and threats can be easily identified on these devices through regular scanning. Threats can be eliminated from the system before any damage can be caused to the system through the use of endpoint security.

- VPN: With the help of VPNs, information/data are transmitted securely over the network from any device. The sensitive data are safely transmitted in the encrypted form over a secured tunnel, which prevents any form of unauthorized access due to eavesdropping.

- Secure web gateway: Secure web gateways are capable of a priori identifying attacks and are crucial for securing cloud services. Cloud security operates on the DNS and IP layers and can recognize phishing, malware, and ransomware attacks. The integration of security with the cloud helps to locate faults on any of the network branches and stops them from spreading onto other branches.

- Email security: Email is one of the most popular media of communication and is more prone to attack. Typically, attackers propagate ransomware and malware through emails. Email security provides advanced threat detection, blocking, and takes necessary corrective actions to avoid data loss and secure information/data using end-to-end encryption.

- Cloud Access Security Broker (CASB): The CASB acts as a tool that operates as a gateway between cloud infrastructure and applications. CASB identifies malicious cloud-based applications and acts as a shield against security breaches.

Summary

1. Safety of professionals and industrial assets at different locations in industries is essential.

2. In order to improve the safety of professionals, certain safety measures are employed, such as the use of appropriate personal safety equipment, operation manuals for machines, safety training at

regular intervals, and strict rules for handling hazardous materials inside a plant.

3. Operational excellence parameters of an organization are comprised of energy consumption, environment, health and safety, quality, operations, and APM. These parameters form the pillars of the organization.

4. IIoT technologies such as smart sensors, AR and VR, protective equipment, and robots are utilized to improve the safety of industrial assets.

5. In order to maintain plant security, the three main pillars of an organization are culture, compliance, and capital.

6. To safeguard the assets of an organization, adhere to the regulation policies, and avoid overflows, software security is necessary. The security requirements of an organization are upgraded with time. Integrity, availability, and authentication form the three essential components of plant security.

7. Mobile device security is necessary because these devices are more prone to malicious attacks. The various components used to secure mobile devices are—endpoint security, VPN, secure web gateway, email security, and CASB.

Exercises

1. What are the safety measures to be taken in a plant by the working personnel?

2. What are the operational excellence parameters? How do these parameters affect the industrial processes?

3. Discuss the safety aspects to be undertaken by workers at workplace in IIoT scenario.

4. What are the main pillars of plant security in an organization? Discuss.

5. What are the essential requirements to provide software security?

6. What is the primary objective for providing network security?

7. Discuss (a) behavioral analytics, (b) antivirus and anti-malware software, (c) web security, and (d) IPS.

8. Why is security of mobile devices necessary?

9. How to secure mobile devices using email security, secure web gateway, and endpoint security?

10. Discuss the security aspects of various IIoT applications in a plant.

17

Case Studies

Learning Outcomes

- ■ This chapter provides an introduction to the various use-cases of IIoT in real-world scenarios.

- ■ New readers will be able to have a grasp of the various types of sectors in industrial settings, which are possible users of this technology.

- ■ Seasoned learners will be able to appreciate the speed with which IIoT and Industry 4.0 are being adopted in the real-world.

17.1 Introduction

IIoT is revolutionizing the industrial landscape world-over. The use of new and legacy technologies, upgrading particular facets of industries, factory floors, and plants to make whole organizations and enterprises compliant to the Industry 4.0 standards has resulted in significant benefits. These benefits include increased reliabilities, efficiencies, power savings, worker safety, security, and profits. In this chapter, we present some of the motivating use-cases of the use of IoT in various industries around the world. We discuss how these industries used IoT to modernize their infrastructure and enhance the manageability of these organizations. We take the example of three widely separated industries, where IIoT has made significant differences in their profits, operations management, or productivity.

17.2   Manufacturing Industry

Manufacturing industries have to rely on proper and coordinated actions/operations to enhance their productivity. All the while, there is a constant need for quality checks at various stages of the manufacturing line, and ensuring quality control of finished products by such industries.

17.2.1   Background of the industry

Black and Decker are globally known for manufacturing power tools as their main product. Focusing on the rapid industrial penetration of IoT for solving some of the long-standing challenges in manufacturing, they approached Cisco and Aeroscout Industrial for enabling connected enterprise-level visibility solutions for their manufacturing plant in Mexico.

17.2.2   Challenges

The need for increased visibility in manufacturing processes is a significant challenge. Additionally, the need for a decrease in manufacturing complexities further adds to the challenges of such a scenario. The constant need for ensuring the quality of products leaving the production/assembly line requires well-connected and orchestrated solutions.

17.2.3   Industrial IoT as a solution

Every material in the manufacturing plant was attached with a WiFi-based RFID tag to monitor their location in real-time continuously. Cisco's wireless network solutions and Aeroscout's tags communicated to a PLC-based plant controller. The controller tracks every stage of the production line and regulates its operations. The use of real-time tracking enabled the floor managers to change the speeds of various sections of the production line and monitor the worker's productivity.

17.2.4   Benefits

The following benefits resulted directly as a result of this solution:

(a) Worker productivity increased by 10%.

(b) Oversight of defects reduced by 16%.

(c) Utilization rates of labor-critical resources improved by 10%.

17.3 Automotive Industry

Automotive industries are made up of a complex linking of various sub-units, which typically deal with one or more components of the finished automobile. Starting from material procurement to designing, engine and mechanism production, electrical and electronics, assembly lines, and sales, every aspect of the automobile industry works in a closely coordinated manner to estimate the demand-supply gap and regulate their production.

17.3.1 Background of the industry

The Hirotec Group of Japan is one of the globally renowned private automotive component manufacturers. Focusing on adopting Industry 4.0 standards to cut-back on unplanned downtimes, Hirotec approached Hewlett Packard Enterprise (HPE) for enabling IoT and IoT-enabled predictive analytics for their manufacturing line.

17.3.2 Challenges

Given Hirotec's ideologies, the most persistent issues in their manufacturing plants were dealing with issues of minimizing unplanned downtimes and ensuring continuity of plant operations.

17.3.3 Industrial IoT as a solution

HPE's IoT framework, paired with HPE Edgeline systems, was piloted in three of Hirotec's facilities.

(a) The first pilot installation in their Detroit plant collected data from CNC machines. Eight CNC machines were integrated with the IoT framework for data collection.

(b) The second pilot was used for monitoring an automated exhaust system inspection line, which included components such as robots, cameras, force sensors, and measurement devices. The system provided remote visualization of the inspection line through the data gathered from all these components.

(c) The third pilot is focusing on the automobile door production facility of Hirotec. The system is expected to provide real-time visualizations and generate automatic reports on the production line operations and quality.

17.3.4   Benefits

The company managed to enhance its productivity and efficiency using the Industry 4.0 framework. The use of predictive analytics on real-time data also enabled them to predict failures, bottlenecks in the production line, and estimate faults.

17.4   Mining Industry

As of 2019, mining is a globally USD 689 billion industry. Mining has many facets and varies significantly depending on the material/substance being mined. The setup, standards, and requirements for open mines are quite different from underground mines. However, almost all mining industries rely on conveyer-belt based systems for loading/unloading of mined materials. This alone makes the production, installation, and maintenance of conveyer belt systems a crucial aspect of the mining industry. Any downtime on the conveyor belts has the potential to impact the mining output directly.

17.4.1   Background of the industry

Transco Industries, a manufacturer of conveyer belt systems for mining industries, is also responsible for their commissioning and upkeep. The company focused on IoT-based solutions for reducing failures and maintenance costs of their installations in and along with a mining setup at Salt lake City, Utah, USA.

17.4.2   Challenges

Typically, these systems in mining can be as long as a few miles. Each subsection of these conveyer belt systems has various pulleys, gears, and bearing mechanisms, which need continuous monitoring for changes in temperature, pressure, and other factors. The upkeep of these subparts of the conveyor system is crucial in enhancing the uptime of the conveyor belts.

17.4.3   Industrial IoT as a solution

Transco used Semtech's LoRa devices and LoRaWAN protocol to monitor their conveyer belt systems. The long communication range of LoRa, along with low power consumption, geolocation, and the advantage of being an open standard, made it the primary choice for monitoring the conveyer belt systems. This setup enables them to monitor defects, abnormal temperature changes, and other factors, which may hinder the smooth operations of the conveyer belt system or may lead to downtimes.

17.4.4 Benefits

The use of this technology has enabled them to have a holistic view of their whole conveyer belt line simultaneously, localize defects, and reduce human errors. This has not only reduced the maintenance costs but also enhanced the profits due to the use of less labor-intensive solutions for monitoring large stretches of the belt systems.

Summary

This chapter provided a brief overview of the real-world use of IoT in industrial environments. The incorporation of IIoT is proving beneficial not only to regular companies, which regularly handle data but also for companies, which have never explored such solutions. The use-cases also indicate significant and multi-faceted benefits for these companies. This chapter aims to motivate the reader about the usefulness of IIoT and Industry 4.0 and emphasize that the present-day industries have already moved to the adoption of such technologies, which are rapidly becoming the new normal.

18

Test Your Understanding

In this chapter, we outline some conceptual problems in the context of IIoT and Industry 4.0. These questions are designed to utilize the reader's understanding of concepts while identifying possible solutions to problems. The readers will have to consult external topic-specific resources, which may be available as books or online resources while answering these questions.

1. What is a transducer? Can actuators be identified as transducers?

2. How do piezoelectric materials work? Are they suitable as sensors or actuators?

3. A silicon-based photocell has a current output of 35 mA/cm^2 and an efficiency of 20%. Calculate the total current generated for a series connection of such photocells under full illumination and no external losses. The total area of this array is 1 m^2.

4. A temperature sensor reads $99^{\circ}C$ for an actual temperature value of $99.99^{\circ}C$. Identify the category of error in this sensor. Justify.

5. A pressure sensor **A** reads 55.3 psi, while another sensor **B** reads 54.1 psi for an actual pressure reading of 55.2 psi. Which of the two sensors is more accurate, and by how much?

6. Can pressure sensors be used for detecting sounds?

7. Differentiate between sensor accuracy and precision.

8. What is a strain gauge? What are its applications?

9. What is the principle of operation of capacitive sensors?

10. A single-acting cylinder in an industrial pneumatic actuator has a diameter of 90 mm. The gauge pressure shows 500 kPa (100kPa=1 bar). Calculate the force exerted by the piston of the cylinder.

11. A pneumatic lift using a single-acting cylinder has a bore diameter of 150mm. This press is expected to generate a pressure of at least 3000N for an industrial application. Calculate the minimum required pressure to be exerted on the piston to make this possible.

12. A double-acting pneumatic cylinder has an outstroke given by $F = p\pi\frac{d_1^2}{4}$, where p is the gauge pressure and d_1 is the diameter of the full-bore piston. If d_2 be the diameter of the piston rod, calculate the generated instroke force such that $p = 500$ kPa, $d_1 = 90$ mm, and $d_2 = 5$ mm.

13. Calculate the force exerted from a single-acting pneumatic cylinder with a full-bore diameter of 0.1 m and a rod diameter of 0.05 m at 1 bar pressure.

14. Differentiate between the torque and the force of a motor.

15. How is light intensity measured. What is the unit of its measurement?

16. An industrial motor has a torque of 50 kNm and rotates at a speed of 30 RPS. Calculate the power required by the motor.

17. A mechanical actuator uses the power of 1450 W for its operation. Calculate its Horse-Power (HP).

18. A motor has a power of 500 kW and a speed of 30 RPM. Calculate the resulting torque in kNM for this motor.

19. A typical industrial setup requires the operation of a power lift under intensely sterile environments. The effect of sparks and moisture can be disastrous for the operation of this setup. What kind of actuator would be best suited for such an operation, and why?

20. The inlet of a single-acting linear hydraulic actuator has a pipe of 0.9 cm radius. If water is pumped into the cylinder with a flow rate of 0.5 L/s, what is the velocity of water in the pipe?

21. The inlet of a single-acting linear hydraulic actuator has a pipe of 0.2 m radius. If water is pumped into the cylinder at a specific flow rate, which results in the velocity of the water in the pipe to be at 0.795 m/s, what is the flow-rate of water through the pipe?

22. An industrial network installation requires the data transmission technology to use twisted-pair cables along with couplers, terminators, and device support for 32 devices per network segment. Out of choice between cyclic data transmission or a polling-based approach, which would be a better strategy and why?

23. Explain the utility of Electronic Device Description Language (EDDL)in Foundation Fieldbus.

24. Explain the utility of Field Device Tool and Device Type Management (FDT/DTM) in Profibus PA.

25. Draw the waveform for the Profibus DP signal transmission 1100101010 through RS-485.

26. Draw the waveform for the Profibus PA signal transmission 1100101010 through RS-485.

27. The HART network uses the 4-20mA signal line. How long will it take to transmit 5kB traffic over this network?

28. Two Interbus devices are separated by 250m from each other on the remote bus. How long will a device take to transmit 1MB using this configuration?

29. Three sensors and two actuators are connected in an Interbus loop. Each of these devices is separated by a distance of at least 35 m from each other. Is this network connection stable and sustainable?

30. A Bitbus installation has been designed to have five segments:

    - Segment-1 has 27 devices.
    - Segment-2 has 15 devices.
    - Segment-3 has 42 devices.
    - Segment-4 has 5 devices.
    - Segment-5 has 180 devices connected through repeaters.

    Is this configuration sustainable? Justify.

31. Two Bitbus devices are separated by 800 m. How long will it take to transmit 750 kB of traffic through the Bitbus network?

32. How many general use I/O and how many dedicated use I/O is supported by CC-Link?

33. Which port is used for Modbus TCP?

34. What are the differences between Modbus TCP and RTU modes?

35. A LonWorks powerline transmits 300kB data. Simultaneously, a Lon-Works twisted pair transceiver transmits the same amount of data. How much time does it take to complete the transmission for each case?

36. Consider an Intensive Care Unit (ICU) of a hospital scenario with some critical patients. The health conditions of these patients are required to be monitored continuously. On the other hand, there is a scarcity of the number of specially trained nurses at that moment in the hospital. Therefore, design a system to monitor and care these critical patients.

37. Google, TomTom, Waze, and Hero are different types of organization which provides online services to the customers related with the road congestion, shortest path, and tracking of road. Design a block-diagrammatic architecture for providing safety information to customers.

Consider the presence of emergency vehicles such as ambulance and time-critical data such as accident-related information and natural disaster. [Hint: You may consider edge/fog nodes for the computation of time-critical data.]

38. Consider the features of a smart factory, as described in the chapter. Design a flowchart to show the step-by-step operation of a smart manufacturing industry with smart technologies, robots performing operation to help humans, remotely monitoring the inventory, working personnel remotely monitor the activities of the plant, control the operations and monitor them, automatic packaging of products, and manage the warehouse activities.

39. Consider the basic IoT four-layered architecture and design the architecture of a smart home.

40. MQTT is a publish-subscribe based application layer protocol. Discuss an application-specific case study in the IoT scenario, where MQTT protocol is used.

41. Based on the services provided by the Cloud Service Provider (CSP), there are three types of cloud – Infrastructure-as-a-Service (IaaS), Platform-as-a-Service (PaaS), and Software-as-a-Service (SaaS). Give practical applications related with each of the type of cloud services.

42. How analytics is associated with the safety of workers in the industries?

43. What are the different workplace distractions which may lead to an accident in industries?

44. Which sensors are generally used in manufacturing industries?

45. Consider the design requirements of a smart business model, and design a business model applicable for software development industries?

46. Design a reference architecture to be applicable for small-scale industries.

47. What are the leading and lagging KPIs for (a) university, (b) super speciality hospital, (c) thermal power plant, and (d) electricity distribution organization?

48. In a special care unit of a hospital, 5 patients possess wearable healthcare systems. Each of these wearable healthcare unit consists of 4 different types of sensors – ECG, pulse oximeter, temperature sensor, and GSR. These sensor nodes transmit the sensed data of the patients via Bluetooth to the local processing unit (LPU). Further, the LPU transmit these data to the cloud using a GSM module.

- Suppose, the ECG sensor transmits data twice per hour, and the other sensors transmit data per minute to the cloud. Estimate the daily wearable healthcare systems to cloud end-to-end overhead for each patient, if heartbeat = 1.85B, other sensor nodes = 1B, the MQTT protocol approximately uses 70B and 85B of data to establish connection between the sensor nodes to LPU and LPU to cloud, respectively. What will be the overhead for the entire hospital unit?

- Estimate the end-to-end overhead over a month (30 days), if the ECG sensor transmits data at an interval of 15 mins per hour and other sensors transmits 6 times per hour. What will be the overhead in terms of KB?

- Consider that the power consumption of each Bluetooth module is 0.035 W and GSM module is 0.05 W for transmission of 100 bytes, find the daily power consumption for

  - each healthcare unit, ECG sensor transmits data once per hour and other sensors twice per hour.
  - 5 patients, when each sensor transmits data twice per hour.

49. An organization, XYZ, performs capacity planning of 5 fog nodes with varying capacity and two-level memory system in each of them. The capacity of fog nodes is 72 KB, 124 KB, 246 KB, 98 KB, and 5 MB respectively. Suppose, the average cost of processing of data packets at the fog nodes be c_1, c_2, c_3, c_4, and c_5 respectively, such that $c_3 = 25c_5$, $c_4 = 15c_1$, $c_2 = 5c_4$, and $c_5 = 12c_1$. Additionally, the number of tasks executed by the fog nodes be p_1, p_2, p_3, p_4, and p_5, respectively. The number of processes executed by the fog node 1 is 5 times greater compared to fog node 3. On the other hand, the number of tasks executed by fog node 2 and 5 is twice that of fog node 1. Moreover, the time required to access the fog nodes is as follows: $t_3 = 10t_2$, $t_5 = 7t_3$, $t_2 = t_1$, and $t_4 = 5t_2$. Each of these fog nodes possess 12%, 15%, 20%, 18%, and 22% of their capacity as cache memory.

   - Calculate the average cost of processing of data packets at the fog nodes, if $c1 = \$0.75$ per KB.

   - Calculate the average number of tasks processed at the fog nodes, if $p_1 = 50$ tasks per min, and $p_4 = 75$ tasks per min.

   - Calculate the time required for processing the data at the fog nodes, if $t_2 = 10$ msec.

   - Estimate the capacity of each of the nodes and compare them based on their capacity, average cost of processing the data packets, number of tasks processed, time required to process, and cache memory.

50. Consider an Industry 4.0 compliant manufacturing industry, where heterogeneous type of sensor nodes [213] are implemented at different locations, which gives rise to TB amount of data. In one of the units of the manufacturing industry, 250 sensor nodes of 10 different types are deployed at different locations. Table 18.1 provides the detailed description of the type of sensor nodes.

Table 18.1: Description of node type.

Type of node	Number of data packets generated/sec	Number of data packets successfully delivered/sec	Time required to reach the server (sec) (per 1000 packets)
A	1750	1525	0.15
B	2467	2280	0.24
C	1876	1584	0.55
D	1655	1325	0.65
E	2275	2075	0.45
F	1525	1415	0.32
G	3320	3175	0.18
H	2375	2125	0.21
I	1835	1450	0.19
J	2150	1735	0.25

- Find the percentage of data packets dropped by each type of node.
- Find the rate at which data packets reach the server by each type of node.
- Estimate the total number of data packets dropped in a day by all the nodes.
- Estimate the average rate at which data packets are dropped by all the nodes.

51. Consider a semi-automated bottle manufacturing industry, where sensor nodes are implemented at different locations. Each of the sensor consists of a sensing and transmission unit. The power required to transmit a bit in Zigbee is 220 uW/bit. An implementation has 250 Zigbee slave nodes (where each node is attached with five sensors), each of which periodically transmits data to a master node at an interval of 60 seconds. The packet from each node consists of a 64-bit node identifier, and a 64-bit destination identifier besides the 8-bit sensor values and 8-bit sensor

identifier from each sensor. How much will be the power consumed by the implementation (except at the master node) for 2-hour of operation.

52. Consider an Intelligent Transportation system, where Safety-as-a-Service (Safe-aaS) [211] architecture is implemented. A Safe-aaS architecture provides customized safety-related decisions dynamically to the end-users. In the Safe-aaS infrastructure, five layers are present – device, edge, decision, decision virtualization, and application layer. Different type of sensor nodes are deployed in the device layer of Safe-aaS. Based on the packet delivery ratio, average state, and utility of a sensor node, it is selected for operation. Table 18.2 provides some specifications. The seven types of sensor nodes are as follows:

Table 18.2: Description of 7 type of nodes.

Type of node	A	B	C	D	E	F	G
Packets sent	3055	3265	2875	3245	2675	3565	2975
Packets dropped	125	137	152	135	215	225	175
Energy utilized for sensing	0.005	0.0065	0.0045	0.0053	0.0062	0.007	0.0067
Energy required for transmission	0.006	0.0055	0.0052	0.0045	0.0033	0.0037	0.0025

- Estimate the packet delivery ratio of the different type of sensor nodes, if 250 sensor nodes are present in the scenario.
- Considering that distance between the sensor nodes and edge nodes, and communication range of these sensor nodes are constant, compute the utility of the sensor nodes. [Hint: Packet delivery ratio =

(Packets sent - Packets dropped)/(Packets sent), and Utility = (Energy required for sensing)/(Energy required for transmission)X(Residual energy X communication range)/(distance X initial energy). Residual energy = Initial energy − (energy required for sensing + energy required for transmission), distance = 115 m, communication range = 220 m, initial energy = 0.5 J]

53. Typically, a Safe-aaS architecture comprises four different actors – end-users, safety service provider (SSP), sensor owner, and vehicle owner [211]. The end-users provide rent to a SSP and enjoys services. On the other hand, the sensor and vehicle owner rent their sensor nodes to the infrastructure. Therefore, business transaction takes place among the different actors [213]. The rent paid by the end-users depend upon various factors such as number of decision parameters selected, time of use, discount given, penalty cost, time when service is received, and allowable time duration for service. Table 18.3 provides some information of the description of decision parameters selected by end-users.

Table 18.3: Description of decision parameters, service duration selected by end-user.

End user	No of decision parameters	Duration (hrs)	Service received
1	5	4	0.77
2	7	5	0.85
3	6	5	1
4	8	6	0.75
5	4	7	0.89
6	5	6	1.1
7	4	5	0.79
8	7	4	0.85
9	8	6	0.9

Discount given = $2, Cost of each template = $5, allowable time duration = 0.75sec, penalty cost = $3.75, t = 5.75 hr, find the amount of penalty paid by SSP, discount received, and amount paid by each end-user. [Hint: A penalty amount is paid by the SSP when time decision is received greater than the allowable time delay (t). Discount is received by end-user, if the services are requested by end-users beyond the specified time duration. Amount paid by an end-user = (cost of each template × number of decision parameters × duration of service) - penalty - discount.]

54. In a Safe-aaS architecture, different type of sensor nodes such as static, inbuilt, and externally placed sensor nodes are present [211, 212]. Static sensor nodes are placed in a particular geographical location. The inbuilt type sensor nodes are present into the vehicles during manufacturing, whereas the externally placed type sensor nodes are externally placed into the vehicles. Suppose, these sensor nodes are activated at certain interval. Table 18.4 provides some data of the service duration and number of activated sensor nodes.

Table 18.4: Description of service duration and activated sensor nodes.

Time duration	No of activated sensor nodes
$T1$	250
$T2$	275
$T3$	285
$T4$	325
$T5$	350
$T6$	290
$T7$	310
$T8$	260
$T9$	290

- Estimate the percentage of activated sensor nodes, if total number of sensor nodes are 400. Additionally, the number of activated inbuilt, static, and externally placed type vary as 35%, 50%, 15% of total activated sensor nodes. Find the total number of activated sensor nodes of each type.

- Estimate the number of static, inbuilt, and externally placed type sensor nodes, if the total number of sensor nodes is 500, and their percentage vary as 60%, 20%, and 20% respectively.

55. Mean time between failure (MTBF) of a smart sensor is referred to the average amount of time that the sensor node functions or operates without failure. Therefore, reliability of the sensor node is described in terms of MTBF. It is measured in terms of operational hours per failure.

- We consider a smart transportation environment, where different types of sensor nodes are deployed at various locations on the road and road side units (RSU). Let us consider an area of 50×50 km^2. If 10,000 sensor nodes are deployed, each with MTBF of 1,85,00,000 hours. Estimate the time in days when the sensor node will fail.

- Suppose the battery life of each sensor node is 125 hours. If 5000 of these sensor nodes are activated every hour and others are inactive, then estimate the time in days when these sensor nodes need to be replaced. What is the MTBF of the battery of these nodes?

- The sensor nodes are manufactured by an organization, ABC. Based on their type, there are certain specifications of these sensor nodes, which are mentioned by the organization. For a specific sensor node, the warranty period given by the manufacturer is 75 days. What would you conclude about the failure of these sensor node?

56. In a thermal power plant, the heat energy is converted to electric power. The water is heated in the boiler and passed through the superheater to convert the saturated steam into superheated steam. Further, this superheated steam is used in steam turbines for generation of electricity.

    - Suppose, a turbine rotates at 7200 RPM. Compute the average rotational latency of the turbine.

    - Different type of sensor nodes are deployed at various locations in the plant. If each sensor node consumes energy of 0.012 mW-h and initial energy is 0.5 J, find the average power consumed per day for 1080 nodes in that unit. Also estimate the average yearly power consumption based on the type of sensor nodes, in the year 2024 (Table 18.5).

Table 18.5: Description of type of sensor nodes and energy consumed.

Type of sensor nodes	Energy consumed (J)
A	0.012
B	0.015
C	0.011
D	0.013
E	0.009
F	0.0115
G	0.0085
H	0.016
I	0.014
J	0.0075

57. In a garment industry, 15 different products are packed in 7 units. Various sensors are attached at various locations of the plant. The entire

process takes place with minimum or no human intervention. Suppose, I – Wrapping and Compression, II – Filling with goods for protection, and III – Cartooning, applying adhesive, and bar-coding, are the processes involved in packaging the garments. Table 18.6 provides details of the units, products, time required, and sub-processes involved.

Table 18.6: Description of units, products, time required, and sub-processes involved.

Units	Products	Time required (sec)	Sub-processes
Unit 1	3, 4	15, 12	I, II
Unit 2	1, 5, 7	11, 17, 13	I, II, III
Unit 3	2, 6	12, 10	I, II
Unit 4	8, 11	16, 13	II, III
Unit 5	9, 10	11, 15	I, II
Unit 6	12, 15	11, 17	II, III
Unit 7	14, 13	13, 17	I, II

- Average time required to pack the garments in all the units.
- Consider, the different type of products packaging time is described as: $P_1 = P_3 = P_5$, $P_2 = 2P_3$, $P_4 = 5P_1$, $P_7 = P_6 = 6P_4$, $P_8 = P_{12} = P_{15} = 3P_2$, $P_9 = P_{11} = 2P_{12}$, $P_{13} = P_3$, $P_{15} = P_4$. Given, $P1 = 100$ units/hr, estimate the number of each product produced per hour. Also find the total time required to pack these products.
- If the number of damaged/rejected garments produced per month is 5, calculate the average number of damaged/rejected garments in coming 5 years.

58. In a milk processing industry, various type of food items are processed and packed in different forms. Different sensors are placed at various locations to check the product quality. Availability of a device = MTBF/ (MTBF + MTR), MTBF and MTR represent the mean time between failure and mean time to repair.

- Estimate the availability for each device.
- Calculate the availability of the system.

59. The failure and maintenance of a system involves $45000. The smart devices attached with these products has an operational probability of 0.96. Estimate the expected cost of failure. If another component is

Table 18.7: Description of units, products, time required, and sub-processes involved.

Machine type	MTBF	MTR
I	220	2
II	325	1.5
III	235	2.25
IV	240	1.75
V	315	1.9
VI	320	2
VII	250	1.75
VIII	330	2.5
IX	310	1.85
X	342	2
XI	318	1.75
XII	215	1.85
XIII	235	2.5
XIV	285	1.9
XV	301	2

added to the system of cost \$150, estimate the cost of failure for the additional component. Consider the operational probability is also 0.96.

60. An IoT-based mine declared special safety measures for the workers going inside the mines. Different sensors are attached to the hard hat and shoes so that the workers can be tracked from a distance. Each of the group of workers going inside the mines has a team head to instruct them (Table 18.7).

   - Suppose the operation of the mines is given below in Table 18.8. Consider the team I, II, III, IV, and V consists of 4, 5, 6, 4, and 7 team members. Each person possesses 4 sensor nodes attached to their uniform. The power consumed by each of these sensor nodes are 0.012 W, 0.015 W, 0.0009 W, and 0.00075 W. Compute the average number of sensor nodes operating over the duration of 15 days.

   - Estimate the average daily power consumption of each team.

   - The reliability of these sensor nodes are 0.96, 0.95, 0.98, and 0.97 respectively. What is the probability that a warning is provided?

Table 18.8: Description of the operation of mines.

Day	Team
1	*I, II*
2	*II, III*
3	*I, II, III*
4	*I, III*
5	*III, IV*
6	*IV, V*
7	*II, III*
8	*IV, V*
9	*II, IV*
10	*I, III*
11	*IV, V*
12	*II, IV*
13	*I, III*
14	*I, II, IV*
15	*II, IV, V*

References

[1] LAN/MAN Standards Committee, 2002. IEEE Standard for Local and Metropolitan Area Networks: Overview and Architecture (En línea). New York, NY, USA. The Institute of Electrical and Electronics Engineers Inc.

[2] Forouzan, B. A. and Fegan, S. C., 2002. TCP/IP Protocol suite. McGraw-Hill Higher Education.

[3] Wilder, F., 1998. A guide to the TCP/IP Protocol Suite. Artech House, Inc.

[4] Popescu-Zeletin, R., 1983, October. Implementing the ISO-OSI reference model. In ACM SIGCOMM Computer Communication Review (Vol. 13, No. 4, pp. 56–66). ACM.

[5] TCP/IP Fundamentals for Windows. Available online: https://technet.microsoft.com/en-us/library/bb726993.aspx

[6] Da Xu, L., He, W. and Li, S., 2014. Internet of things in industries: A survey. IEEE Transactions on industrial informatics, 10(4), pp. 2233-2243.

[7] Vermesan, O. and Friess, P., 2013. Internet of Things: converging technologies for smart environments and integrated ecosystems. River Publishers.

[8] Singh, M., Rajan, M.A., Shivraj, V.L. and Balamuralidhar, P., 2015, April. Secure MQTT for internet of things (IoT). In 2015 Fifth International Conference on Communication Systems and Network Technologies (pp. 746-751). IEEE.

[9] MQTT. Available online: http://mqtt.org/

[10] CoAP. Available online: https://tools.ietf.org/html/rfc7252

[11] AMQP. Available online: https://www.amqp.org/about/what

[12] RFC. Available online: https://tools.ietf.org/html/rfc6455

[13] 6LoWPAN. Available online: https://tools.ietf.org/html/rfc6282

[14] IPv6. Available online: https://www.internetsociety.org/deploy360/ipv6/

[15] RPL. Available online: https://tools.ietf.org/html/rfc6550

[16] Bluetooth. Available online: https://learn.sparkfun.com/tutorials/bluetooth-basics

[17] WiFi. Available online: https://www.tutorialspoint.com/wi-fi/

[18] IoT Protocols. Available online: https://www.postscapes.com/internet-of-things-protocols/

[19] Sehgal N., and Bhatt P. C. P., 2018. Cloud Computing. Springer International Publishing.

[20] Al-Fuqaha, A., Guizani, M., Mohammadi, M., Aledhari, M. and Ayyash, M., 2015. Internet of things: A survey on enabling technologies, protocols, and applications. IEEE communications surveys & tutorials, 17(4), pp. 2347-2376.

[21] Fog Computing. Available online: https://blogs.cisco.com/perspectives/iot-from-cloud-to-fog-computing

[22] Bonomi, F., Milito, R., Zhu, J. and Addepalli, S., 2012, August. Fog computing and its role in the internet of things. In Proceedings of the first edition of the MCC workshop on Mobile cloud computing (pp. 13-16). ACM.

[23] Youseff, L., Butrico, M. and Da Silva, D., 2008, November. Toward a unified ontology of cloud computing. In 2008 Grid Computing Environments Workshop (pp. 1-10). IEEE.

[24] Marston, S., Li, Z., Bandyopadhyay, S., Zhang, J. and Ghalsasi, A., 2011. Cloud computing—The business perspective. Decision support systems, 51(1), pp. 176-189.

[25] Vaquero, L.M., Rodero-Merino, L., Caceres, J. and Lindner, M., 2008. A break in the clouds: towards a cloud definition. ACM SIGCOMM Computer Communication Review, 39(1), pp. 50-55.

[26] Leimeister, S., Böhm, M., Riedl, C. and Krcmar, H., 2010. The business perspective of cloud computing: actors, roles and value networks.

[27] Virtualization in Cloud Computing. Available online: https://www.javatpoint.com/virtualization-in-cloud-computing

[28] Types of Virtualization. Available online: https://redswitches.com/blog/different-types-virtualization-cloud-computing-explained

[29] Liu, F., Tong, J., Mao, J., Bohn, R., Messina, J., Badger, L. and Leaf, D., 2011. NIST cloud computing reference architecture. NIST special publication, 500(2011), pp. 1-28.

[30] Alzahrani, A., Alalwan, N. and Sarrab, M., 2014, April. Mobile cloud computing: advantage, disadvantage and open challenge. In Proceedings

of the 7th Euro American Conference on Telematics and Information Systems (p. 21). ACM.

[31] Yi, S., Li, C. and Li, Q., 2015, June. A survey of fog computing: concepts, applications and issues. In Proceedings of the 2015 workshop on mobile big data (pp. 37-42). ACM.

[32] Bonomi, F., Milito, R., Natarajan, P. and Zhu, J., 2014. Fog computing: A platform for internet of things and analytics. In Big data and internet of things: A roadmap for smart environments (pp. 169-186). Springer, Cham.

[33] Stojmenovic, I., 2014, November. Fog computing: A cloud to the ground support for smart things and machine-to-machine networks. In 2014 Australasian Telecommunication Networks and Applications Conference (ATNAC) (pp. 117-122). IEEE.

[34] Yuriyama, M. and Kushida, T., 2010. Sensor-cloud infrastructure-physical sensor management with virtualized sensors on cloud computing. NBiS, 10, pp. 1-8.

[35] Misra, S., Chatterjee, S. and Obaidat, M.S., 2014. On theoretical modeling of sensor cloud: A paradigm shift from wireless sensor network. IEEE Systems journal, 11(2), pp. 1084-1093.

[36] Jajodia, S., Noel, S. and O'berry, B., 2005. Topological analysis of network attack vulnerability. In Managing Cyber Threats (pp. 247-266). Springer, Boston, MA.

[37] Industry X.0. Available online: https://www.accenture.com/us-en/services/industryx0-index

[38] IIoT Market research. Available online: https://www.globenewswire.com/news-release/2018/05/28/1512669/0/en/Global-Industrial-Internet-of-Things-IIoT-Market-Size-Will-Grow-232-15-Bn-by-2023-Zion-Market-Research.html

[39] Brynjolfsson, E. and McAfee, A., 2014. The second machine age: Work, progress, and prosperity in a time of brilliant technologies. WW Norton & Company.

[40] Industrial Internet. Available online: https://www.genewsroom.com/sites/default/files/media/201306/5901.pdf

[41] Schwab, K., 2017. The fourth industrial revolution. Crown Publishing Group.

[42] Industrial Internet. Available online: https://www.ge.com/docs/chapters/Industrial-Internet.pdf

[43] Lu, Y., 2017. Industry 4.0: A survey on technologies, applications and open research issues. Journal of Industrial Information Integration, 6, pp. 1-10.

[44] Industry 4.0. Available online: http://www.cannonautomata-applications.com/industry40/

[45] Industry 4.0 and IIoT. Available online: https://www.automationworld.com/industry-40-or-industrial-internet-things-whats-your-preference

[46] Jeschke, S., Brecher, C., Meisen, T., Özdemir, D. and Eschert, T., 2017. Industrial internet of things and cyber manufacturing systems. In Industrial Internet of Things (pp. 3-19). Springer, Cham.

[47] Industrial IoT. Available online: https://medium.com/iotforall/what-product-managers-need-to-know-about-industrial-iot-8c92eec1d9d2

[48] CPS. Available online: https://www.intel.com.br/content/www/br/pt/education/highered/cyber-physical-systems-video.html

[49] Ma, D., Fan, X., Gausemeier, J., and Grafe, M., 2011. Virtual reality & augmented reality in industry. Springer.

[50] Anghel, A., Vasile, G., Boudon, R., d'Urso, G., Girard, A., Boldo, D. and Bost, V., 2016. Combining spaceborne SAR images with 3D point clouds for infrastructure monitoring applications. ISPRS journal of photogrammetry and remote sensing, 111, pp. 45-61.

[51] Schwab, K., 2017. The fourth industrial revolution. Crown Publishing Group.

[52] Paton, A.W., Morona, R. and Paton, J.C., 2012. Bioengineered microbes in disease therapy. Trends in molecular medicine, 18(7), pp. 417-425.

[53] Industry 4.0 Technologies. Available online: https://ottomotors.com/blog/5-industry-4-0-technologies

[54] Fourth Industrial Revolution. Available online: https://www.ge.com/reports/the-4th-industrial-revolution-what-it-means-how-to-respond/

[55] Garbie, I., 2016. Sustainability in manufacturing enterprises: Concepts, analyses and assessments for industry 4.0. Springer.

[56] Stock, T. and Seliger, G., 2016. Opportunities of sustainable manufacturing in industry 4.0. Procedia Cirp, 40, pp.536-541.

[57] Amit, R. and Zott, C., 2001. Value creation in e-business. Strategic management journal, 22(6-7), pp. 493-520.

[58] Ustundag, A. and Cevikcan, E., 2017. Industry 4.0: managing the digital transformation. Springer.

[59] Smart and Connected Products. Available online: https://www.tcs.com/perspectives/articles/why-your-products-must-be-smart-and-connected

[60] Thames, L. and Schaefer, D., 2017. Cybersecurity for Industry 4.0: Analysis for Design and Manufacturing. Springer Publishing Company, Incorporated, 1st edition.

[61] Thames, L. and Schaefer, D., 2016. Software-defined cloud manufacturing for industry 4.0. Procedia cirp, 52, pp. 12-17.

[62] Fulcher, J., 2008. Computational intelligence: an introduction. In Computational intelligence: a compendium (pp. 3-78). Springer, Berlin, Heidelberg.

[63] Boyes, H., Hallaq, B., Cunningham, J. and Watson, T., 2018. The industrial internet of things (IIoT): An analysis framework. Computers in Industry, 101, pp.1-12.

[64] IoT Architecture. Available online: https://medium.com/datadriveninvestor/4-stages-of-iot-architecture-explained-in-simple-words-b2ea8b4f777f

[65] IoT. Available online: https://info.rti.com/hubfs/docs/Industrial IoT-FAQ.pdf

[66] IIC. Available online: https://www.iiconsortium.org/about-us.html

[67] Industrial Internet. Available online: http://wikibon.org/wiki/v/DefiningandSizingtheIndustrialInternet

[68] Industrial Internet. Available online: https://www.ge.com/docs/chapters/IndustrialInternet.pdf

[69] Industrial Automation. Available online: https://www.surecontrols.com/what-is-industrial-automation/

[70] Smart Sensors. Available online: https://www.mepits.com/tutorial/193/basic-electronics/smart-sensors

[71] Available online: https://www.azosensors.com/article.aspx?ArticleID=1289

[72] Industry 4.0. Available online: https://www.pwc.com/gx/en/industries/industries-4.0/landing-page/industry-4.0-building-your-digital-enterprise-april-2016.pdf

[73] Islam, T., Mukhopadhyay, S.C. and Suryadevara, N.K., 2016. Smart sensors and internet of things: a postgraduate paper. IEEE Sensors Journal, 17(3), pp. 577-584.

[74] Lee, J., Bagheri, B. and Kao, H.A., 2015. A cyber-physical systems architecture for industry 4.0-based manufacturing systems. Manufacturing letters, 3, pp. 18-23.

[75] Aazam, M., Zeadally, S. and Harras, K.A., 2018. Deploying fog computing in industrial internet of things and industry 4.0. IEEE Transactions on Industrial Informatics, 14(10), pp. 4674-4682.

[76] IoT Reference Architecture. Available online: Microsoft-Azure-IoT-Reference-Architecture.pdf

[77] IIoT Reference Architecture. Available online: https://www. iiconsortium.org/IIC-PUB-G1-V1.80-2017-01-31.pdf

[78] Digital Enterprise. Available online: https://new.siemens.com/global/en /company/topic-areas/future-of-manufacturing/digitalenterprise.html

[79] Systra Automated Transport. Available online: https://www.systra. com/IMG/pdf/systra-automated-and-autonomous-public-transport.pdf

[80] Osterwalder A., Pigneur Y., and Clark T, 2010. Business Model Generation: A Handbook for Visionaries, Game Changers, and Challengers. Hoboken, NJ: Wiley.

[81] Microsoft Azure. Available online: Microsoft-Azure-IoT-Reference-Architecture.pdf

[82] IIC. Available online: https://www.iiconsortium.org/IIC-PUB-G1-V1.80-2017-01-31.pdf

[83] Reference Architecture. Available online: http://www.calimar.com/ TheConceptOfReferenceArchitectures.pdf

[84] IoT Business Models. Available online: https://www.softwebsolutions. com/resources/how-to-monetize-IoT-business-models.html

[85] KPI. Available online: https://www.klipfolio.com/resources/articles/ what-is-a-key-performance-indicator

[86] Leading/Lagging Indicators. Available online: https://www. leadingagile.com/2018/02/leading-lagging-indicators/

[87] IIoT. Available online: https://www.trendmicro.com/vinfo/us/security /definition/industrial-internet-of-things-iiot

[88] Mubeen, S., Asadollah, S.A., Papadopoulos, A.V., Ashjaei, M., Pei-Breivold, H. and Behnam, M., 2017. Management of service level agreements for cloud services in iot: A systematic mapping study. IEEE Access, 6, pp. 30184-30207.

[89] Papadopoulos, A.V., Asadollah, S.A., Ashjaei, M., Mubeen, S., Pei-Breivold, H. and Behnam, M., 2017, August. SLAs for industrial IoT:

Mind the gap. In 2017 5th International Conference on Future Internet of Things and Cloud Workshops (FiCloudW) (pp. 75-78). IEEE.

[90] IIoT Cloud Platforms. Available online: https://fr.farnell.com/will-there-be-a-dominant-iiot-cloud-platform

[91] Cloud. Available online: https://nvlpubs.nist.gov/nistpubs/Legacy/SP/nistspecialpublication500-292.pdf

[92] Selecting Cloud Platforms. Available online: https://iotify.io/top-10-selection-criteria-for-your-iot-cloud-platform/

[93] Bonomi, F., Milito, R., Zhu, J. and Addepalli, S., 2012, August. Fog computing and its role in the internet of things. In Proceedings of the first edition of the MCC workshop on Mobile cloud computing (pp. 13-16). ACM.

[94] Aazam, M., Zeadally, S. and Harras, K.A., 2018. Deploying fog computing in industrial internet of things and industry 4.0. IEEE Transactions on Industrial Informatics, 14(10), pp. 4674-4682.

[95] GE. Available online: https://www.ge.com/digital/iiot-platform

[96] Mindsphere. Available online: https://new.siemens.com/global/en/products/software/mindsphere.html

[97] Honeywell. Available online: https://www.honeywellprocess.com/en-US/onlinecampaigns/connectedplant/Pages/home.html

[98] C3AI. Available online: https://c3.ai/products/ai-appliance/

[99] Meshify. Available online: https://meshify.com/

[100] Uptake. Available online: https://www.uptake.com/applications

[101] Edge Computing. Available online: https://www.iiconsortium.org/pdf/IntroductiontoEdgeComputinginIIoT2018-06-18.pdf

[102] Fog computing. Available online: https://www.controleng.com/articles/fog-computing-for-industrial-automation/

[103] Yanez, F., 2017. The 20 Key Technologies of Industry 4.0 and Smart Factories: The Road to the Digital Factory of the Future. New York.

[104] Augmented Reality. Available online: https://www.realitytechnologies.com/augmented-reality/

[105] Augmented Reality. Available online: https://en.wikipedia.org/wiki/Augmentedreality

[106] Comport, A.I., Marchand, E., Pressigout, M. and Chaumette, F., 2006. Real-time markerless tracking for augmented reality: the virtual visual servoing framework. IEEE Transactions on visualization and computer graphics, 12(4), pp.615-628.

[107] Comport, A.I., Marchand, É. and Chaumette, F., 2003, October. A real-time tracker for markerless augmented reality. In Proceedings of the 2nd IEEE/ACM International Symposium on Mixed and Augmented Reality (p. 36). IEEE Computer Society.

[108] Augmented Reality. Available online: https://lightguidesys.com/blog/projector-based-augmented-reality-new-form-enterprise-ar/

[109] Virtual Reality. Available online: https://en.wikipedia.org/wiki/ Virtualreality

[110] Obeysekare, U., Williams, C., Durbin, J., Rosenblum, L., Rosenberg, R., Grinstein, F., Ramamurti, R., Landsberg, A. and Sandberg, W., 1996, October. Virtual workbench-a non-immersive virtual environment for visualizing and interacting with 3D objects for scientific visualization. In Proceedings of Seventh Annual IEEE Visualization'96 (pp. 345-349). IEEE.

[111] Bimber, O., Grundhofer, A., Zeidler, T., Danch, D. and Kapakos, P., 2006, March. Compensating indirect scattering for immersive and semi-immersive projection displays. In IEEE Virtual Reality Conference (VR 2006) (pp. 151-158). IEEE.

[112] Applications of AR. Available online: https://www.lifewire.com/applications-of-augmented-reality-2495561

[113] Applications of AR. Available online: https://www.forbes.com/sites/bernardmarr/2018/07/30/9-powerful-real-world-applications-of-augmented-reality-ar-today/3b02cf0c2fe9

[114] Virtual Reality. Available online: https://www.vrs.org.uk/

[115] Big Data. Available online: https://www.ibm.com/analytics/hadoop/big-data-analytics

[116] Big Data. Available online: https://en.wikipedia.org/wiki/Bigdata

[117] Structured vs. Unstructured Data. Available online: https://www.geeksforgeeks.org/difference-between-structured-semi-structured-and-unstructured-data/

[118] Structured vs. Unstructured Data. Available online: https://solutionsreview.com/data-management/key-differences-betw2016.een-structured-and-unstructured-data/

[119] Structured vs. Unstructured Data. Available online: https://www.datamation.com/big-data/structured-vs-unstructured-data.html

[120] Big Data. Available online: https://en.wikipedia.org/wiki/Bigdata

[121] Haykin, S., Wright, S. and Bengio, Y., 2015. Big Data: Theoretical Aspects [scanning the issue]. Proceedings of the IEEE, 104(1), pp. 8-10.

[122] Birney, E., 2012. The making of ENCODE: lessons for big-data projects. Nature, 489(7414), p. 49.

[123] Lyko, K., Nitzschke, M., and Ngonga Ngomo, A. C., 2016. Big Data Acquisition. Springer, Cham

[124] Big Data. Available online: https://www.scnsoft.com/blog/what-is-big-data

[125] Available online: https://www.allerin.com/blog/top-5-sources-of-big-data

[126] Smart Factory. Available online: https://www2.deloitte.com/content/dam /insights/us/articles/4051The-smart-factory/DUPThe-smart-factory

[127] Smart Manufacturing. Available online: https://www.gemalto. com/iot/inspired/smart-manufacturing

[128] JIIT. Available online: https://www.ifm.eng.cam.ac.uk/research/dstools/ jit-just-in-time-manufacturing/

[129] Lean Manufacturing. Available online: https://www.muellercpa. com/newsletters/eight-steps-lean-manufacturing-approach

[130] Value Stream Mapping. Available online: https://www.plutora.com/ blog/value-stream-mapping

[131] Mrugalska, B. and Wyrwicka, M.K., 2017. Towards lean production in industry 4.0. Procedia Engineering, 182, pp. 466-473.

[132] Sanders, A., Elangeswaran, C. and Wulfsberg, J.P., 2016. Industry 4.0 implies lean manufacturing: Research activities in industry 4.0 function as enablers for lean manufacturing. Journal of Industrial Engineering and Management (JIEM), 9(3), pp. 811-833.

[133] Lean Manufacturing. Available online: https://www. processexcellencenetwork.com/lean-six-sigma-business-performance/articles/13-ideas-for-successful-implementation-of-lean-ma

[134] Krishnamurthy, L., Adler, R., Buonadonna, P., Chhabra, J., Flanigan, M., Kushalnagar, N., Nachman, L. and Yarvis, M., 2005, November. Design and deployment of industrial sensor networks: experiences from a semiconductor plant and the north sea. In Proceedings of the 3rd international conference on Embedded networked sensor systems (pp. 64-75). ACM.

[135] Lasi, H., Fettke, P., Kemper, H.G., Feld, T. and Hoffmann, M., 2014. Industry 4.0. Business & information systems engineering, 6(4), pp. 239-242.

[136] Jazdi, N., 2014, May. Cyber physical systems in the context of Industry 4.0. In 2014 IEEE international conference on automation, quality and testing, robotics (pp. 1-4). IEEE.

[137] Soloman, S., 2009. Sensors handbook. McGraw-Hill, Inc..

[138] Van Herwaarden, A.W. and Sarro, P.M., 1986. Thermal sensors based on the Seebeck effect. Sensors and Actuators, 10(3-4), pp.321-346.

[139] Meijer, G.C., 1986. Thermal sensors based on transistors. Sensors and Actuators, 10(1-2), pp. 103-125.

[140] Bao, M.H., 2000. Micro mechanical transducers: pressure sensors, accelerometers and gyroscopes (Vol. 8). Elsevier.

[141] Kong, J., Franklin, N.R., Zhou, C., Chapline, M.G., Peng, S., Cho, K. and Dai, H., 2000. Nanotube molecular wires as chemical sensors. science, 287(5453), pp.622-625.

[142] Kajima, T. and Kawamura, Y., 1995. Development of a high-speed solenoid valve: Investigation of solenoids. IEEE Transactions on industrial electronics, 42(1), pp.1-8.

[143] Stoeckel, D. and Waram, T., 1992, January. Use of Ni-Ti shape memory alloys for thermal sensor-actuators. In Active and adaptive optical components (Vol. 1543, pp. 382-387). International Society for Optics and Photonics.

[144] Copic, D. and Hart, A.J., 2015. Corrugated paraffin nanocomposite films as large stroke thermal actuators and self-activating thermal interfaces. ACS applied materials & interfaces, 7(15), pp.8218-8224.

[145] Bobrow, J.E. and Lum, K., 1995, June. Adaptive, high bandwidth control of a hydraulic actuator. In Proceedings of 1995 American Control Conference-ACC'95 (Vol. 1, pp. 71-75). IEEE.

[146] Asl, R.M., Hagh, Y.S., Simani, S. and Handroos, H., 2019. Adaptive square-root unscented Kalman filter: An experimental study of hydraulic actuator state estimation. Mechanical Systems and Signal Processing, 132, pp.670-691.

[147] Andrighetto, P.L., Valdiero, A.C. and Carlotto, L., 2006. Study of the friction behavior in industrial pneumatic actuators. In ABCM Symposium series in mechatronics (Vol. 2, No. 2, pp. 369-376).

[148] Boldea, I. and Nasar, S.A., 1999. Linear electric actuators and generators. IEEE Transactions on Energy Conversion, 14(3), pp.712-717.

[149] Verhappen, I. and Pereira, A., 2008. Foundation Fieldbus. ISA.

[150] Tovar, E. and Vasques, F., 1999. Real-time fieldbus communications using Profibus networks. IEEE transactions on Industrial Electronics, 46(6), pp.1241-1251.

[151] Liu, J., Fang, Y. and Zhang, D., 2007, May. PROFIBUS-DP and HART protocol conversion and the gateway development. In 2007 2nd

IEEE Conference on Industrial Electronics and Applications (pp. 15-20). IEEE.

[152] Ho, I.W.H., North, R.J., Polak, J.W. and Leung, K.K., 2011. Effect of transport models on connectivity of interbus communication networks. Journal of Intelligent Transportation Systems, 15(3), pp.161-178.

[153] Virvalo, T. and Puusaari, P., 1989. Developing intelligent actuator systems: experimenting with Bitbus and the 8096. Microprocessors and Microsystems, 13(4), pp.263-270.

[154] Xie, X., Wei, J., and Wang, T., 2010. Monitoring system for the monorail transport at metal mines based on CC-link field bus. Nonferrous Metals (Mining Section), 4.

[155] Swales, A., 1999. Open modbus/TCP specification. Schneider Electric, 29.

[156] BatiBus – Feldbus for Building Automation, KUNBUS GMBH. Available online: https://www.kunbus.com/batibus.html

[157] Dickmann, G., 2011, April. DigitalSTROM: A centralized PLC topology for home automation and energy management. In 2011 IEEE International Symposium on Power Line Communications and Its Applications (pp. 352-357). IEEE.

[158] Ran, L., Junfeng, W., Haiying, W. and Gechen, L., 2010, October. Design method of CAN BUS network communication structure for electric vehicle. In International Forum on Strategic Technology 2010 (pp. 326-329). IEEE.

[159] Biegacki, S. and VanGompel, D., 1996. The application of DeviceNet in process control. ISA transactions, 35(2), pp. 169-176.

[160] Loy, D., Dietrich, D. and Schweinzer, H.J. eds., 2012. Open control networks: LonWorks/EIA 709 technology. Springer Science & Business Media.

[161] Quang, P.T.A. and Kim, D.S., 2013. Throughput-aware routing for industrial sensor networks: Application to ISA100. 11a. IEEE Transactions on Industrial Informatics, 10(1), pp. 351-363.

[162] Kim, A.N., Hekland, F., Petersen, S. and Doyle, P., 2008, September. When HART goes wireless: Understanding and implementing the WirelessHART standard. In 2008 IEEE International Conference on Emerging Technologies and Factory Automation (pp. 899-907). IEEE.

[163] Augustin, A., Yi, J., Clausen, T. and Townsley, W., 2016. A study of LoRa: Long range & low power networks for the internet of things. Sensors, 16(9), p. 1466.

[164] Ratasuk, R., Vejlgaard, B., Mangalvedhe, N. and Ghosh, A., 2016, April. NB-IoT system for M2M communication. In 2016 IEEE wireless communications and networking conference (pp. 1-5). IEEE.

[165] Yu, H., Coverage extension for IEEE802. 11ah. IEEE-SA Mentor, 802, pp. 1-10.

[166] Groover, M.P., 2007. Automation, production systems, and computer-integrated manufacturing. Prentice Hall Press.

[167] Stouffer, K.A., Falco, J.A. and Scarfone, K.A., 2011. Sp 800-82. guide to industrial control systems (ICS) security: Supervisory control and data acquisition (SCADA) systems, distributed control systems (DCS), and other control system configurations such as programmable logic controllers (PLC).

[168] Ericsson, N., Lennvall, T., Akerberg, J. and Bjorkman, M., 2016, July. Challenges from research to deployment of industrial distributed control systems. In 2016 IEEE 14th International Conference on Industrial Informatics (INDIN) (pp. 68-73). IEEE.

[169] Bayindir, R. and Cetinceviz, Y., 2011. A water pumping control system with a programmable logic controller (PLC) and industrial wireless modules for industrial plants. An experimental setup. ISA transactions, 50(2), pp. 321-328.

[170] Frey, G. and Litz, L., 2000, October. Formal methods in PLC programming. In SMC 2000 conference proceedings. 2000 IEEE international conference on systems, man and cybernetics. 'Cybernetics evolving to systems, humans, organizations, and their complex interactions' (cat. no. 0 (Vol. 4, pp. 2431-2436). IEEE.

[171] Knapp, E.D. and Langill, J.T., 2014. Industrial Network Security: Securing critical infrastructure networks for smart grid, SCADA, and other Industrial Control Systems. Syngress.

[172] Available online: https://internetofthingsagenda.techtarget.com/blog/IoT-Agenda/Why-IIoT-needs-real-time-analytics-now-more-than-ever

[173] Civerchia, F., Bocchino, S., Salvadori, C., Rossi, E., Maggiani, L. and Petracca, M., 2017. Industrial Internet of Things monitoring solution for advanced predictive maintenance applications. Journal of Industrial Information Integration, 7, pp. 4-12.

[174] IoT Data Analytics. Available online: https://www.fingent.com/blog/role-of-data-analytics-in-internet-of-things-iot

[175] IIC Industrial Analytics. Available online: https://www.iiconsortium.org/pdf/IICIndustrialAnalyticsFramework Oct2017.pdf

[176] Turing, A.M., 2009. Computing machinery and intelligence. In Parsing the Turing Test (pp. 23-65). Springer, Dordrecht.

[177] Oracle Analytics Cloud. Available online: https://docs.oracle.com/en/cloud/paas/analytics-cloud/tutorials.html

[178] AI. Available online: https://en.wikipedia.org/wiki/Artificialintelligence

[179] C3AI. Available online: https://c3.ai/products/c3-ai-suite/c3-ai-suite-services-and-capabilities/

[180] H2O. Available online: https://www.h2o.ai/products/h2o/features

[181] Toshiba. Available online: https://www.toshiba-sol.co.jp/pro/satlys/kata/estimation.html

[182] Toshiba. Available online: https://www.toshiba-sol.co.jp/pro/bigdataai/indexj.html

[183] Machine Learning Analytics. Available online: https://www.ge.com/digital/iiot-platform/machine-learning-analytics

[184] AI and ML. Available online: https://www.splunk.com/pdfs/ebooks/5-big-myths-of-ai-and-machine-learning-debunked.pdf

[185] Deep Learning. Available online: https://ww2.mathworks.cn/en/discovery/deep-learning.html

[186] GE Digital . Available online: https://www.ge.com/digital/sites/default/files/downloadassets/Data-Science-Services-from-GE-Digital.pdf

[187] Random Forest. Available online: https://dataaspirant.com/2017/05/22/random-forest-algorithm -machine-learing/

[188] SVM. Available online: http://www.robots.ox.ac.uk/ az/lectures/m-l/lect2.pdf

[189] Random Forest. Available online: https://en.wikipedia.org/wiki/Randomforest

[190] KNN. Available online: https://machinelearningmastery.com/k-nearest-neighbors-for-machine-learning/

[191] Expectation-Maximization. Available online: https://machinelearningmastery.com/expectation-maximization-em-algorithm/

[192] K-Means Clustering. Available online: https://towardsdatascience.com/understanding-k-means-clustering-in-machine-learning-6a6e67336aa1

[193] Mohs, R.C. and Greig, N. H., 2017. Drug discovery and development: Role of basic biological research. Alzheimer's & Dementia: Translational Research & Clinical Interventions, 3(4), pp. 651-657.

[194] Hossain, M.S. and Muhammad, G., 2016. Cloud-assisted industrial internet of things (IIoT)–enabled framework for health monitoring. Computer Networks, 101, pp. 192-202.

[195] Roy, A., Roy, C., Misra, S., Rahulamathavan, Y. and Rajarajan, M., 2018, May. Care: criticality-aware data transmission in CPS-based healthcare systems. In 2018 IEEE International Conference on Communications Workshops (ICC Workshops) (pp. 1-6). IEEE.

[196] Nguyen, H.H., Mirza, F., Naeem, M.A. and Nguyen, M., 2017, April. A review on IoT healthcare monitoring applications and a vision for transforming sensor data into real-time clinical feedback. In 2017 IEEE 21st International Conference on Computer Supported Cooperative Work in Design (CSCWD) (pp. 257-262). IEEE.

[197] IoT Healthcare. Available online: https://www.peerbits.com/blog/internet-of-things-healthcare-applications-benefits-and-challenges.html

[198] Fong, E.M. and Chung, W.Y., 2013. Mobile cloud-computing-based healthcare service by non-contact ECG monitoring. Sensors, 13(12), pp.16451-16473.

[199] Health Analytics. Available online: https://www.izenda.com/what-is-healthcare-analytics/

[200] Yang, Z., Zhou, Q., Lei, L., Zheng, K. and Xiang, W., 2016. An IoT-Cloud based wearable ECG monitoring system for smart healthcare. Journal of medical systems, 40(12), p.286.

[201] Baek, H.J., Chung, G.S., Kim, K.K. and Park, K.S., 2011. A smart health monitoring chair for nonintrusive measurement of biological signals. IEEE transactions on Information Technology in Biomedicine, 16(1), pp.150-158.

[202] Medical Technologies. Available online: https://www.proclinical.com/blogs/2019-2/top-10-new-medical-technologies-of-2019

[203] Available online: https://youtu.be/sl5zEPRkp0U

[204] Amazon. Available online: https://www.aboutamazon.com/amazon-fulfillment/our-fulfillment-centers/why-amazon-warehouses-are-called-fulfillment-centers

[205] RFID. Available online: https://blog.atlasrfidstore.com/active-rfid-vs-passive-rfid

[206] IoT Inventory Management. Available online: https://www.scnsoft.com/blog/iot-for-inventory-management

[207] IIoT Plant Safety. Available online: https://www.engineering.com/AdvancedManufacturing/ArticleID/18873 /Using-IIoT-Connected-Devices-for-Worker-Health-Safety.aspx

[208] Rockwell Automation. Available online: https://literature. rockwellautomation.com/idc/groups/literature/documents /wp/safety-wp037-en-p.pdf

[209] CISCO Network Security. Available online: https://www.cisco. com/c/en/us/products/security/what-is-network-security.html how-network-security-works

[210] CISCO Mobile Security. Available online: https://www.cisco. com/c/en/us/solutions/small-business/resource-center/security/mobile-device-security.html components

[211] C. Roy, A. Roy, S. Misra and J. Maiti, 2018. Safe-aaS: Decision virtualization for effecting safety-as-a-service. IEEE Internet of Things Journal, vol. 5, no. 3, pp. 1690-1697.

[212] C. Roy, S. Misra, J. Maiti and M. S. Obaidat. DENSE: Dynamic Edge Node Selection for safety-as-a-service. 2019 IEEE Global Communications Conference (GLOBECOM), Waikoloa, HI, USA, 2019, pp. 1-6, doi: 10.1109/GLOBECOM38437.2019.9014180.

[213] C. Roy, S. Misra and S. Pal, Blockchain-enabled safety-as-a-service for Industrial IoT applications. IEEE Internet of Things Magazine, vol. 3, no. 2, pp. 19-23, June 2020, doi: 10.1109/IOTM.0001.1900080.

Index

Note: Locators in *italics* represent figures and **bold** indicate tables in the text.

Printed in the United States
by Baker & Taylor Publisher Services